SpringerBriefs in Optimization

Series Editors

Panos M. Pardalos
János D. Pintér
Stephen M. Robinson
Tamás Terlaky
My T. Thai

SpringerBriefs in Optimization showcases algorithmic and theoretical techniques, case studies, and applications within the broad-based field of optimization. Manuscripts related to the ever-growing applications of optimization in applied mathematics, engineering, medicine, economics, and other applied sciences are encouraged.

For further volumes:
http://www.springer.com/series/8918

Petros Xanthopoulos • Panos M. Pardalos
Theodore B. Trafalis

Robust Data Mining

 Springer

Petros Xanthopoulos
Department of Industrial Engineering
   and Management Systems
University of Central Florida
Orlando, FL, USA

Theodore B. Trafalis
School of Industrial
   and Systems Engineering
The University of Oklahoma
Norman, OK, USA

School of Meteorology
The University of Oklahoma
Norman, OK, USA

Panos M. Pardalos
Center for Applied Optimization
Department of Industrial
   and Systems Engineering
University of Florida
Gainesville, FL, USA

Laboratory of Algorithms and Technologies
   for Networks Analysis (LATNA)
National Research University
Higher School of Economics
Moscow, Russia

ISSN 2190-8354   ISSN 2191-575X (electronic)
ISBN 978-1-4419-9877-4   ISBN 978-1-4419-9878-1 (eBook)
DOI 10.1007/978-1-4419-9878-1
Springer New York Heidelberg Dordrecht London

Library of Congress Control Number: 2012952105

Mathematics Subject Classification (2010): 90C90, 62H30

Printed on acid-free paper

Springer is part of Springer Science+Business Media (www.springer.com)

To our families for their continuous support on our work...

Preface

Real measurements involve errors and uncertainties. Dealing with data imperfections and imprecisions is one of the modern data mining challenges. The term "robust" has been used by different disciplines such as statistics, computer science, and operations research to describe algorithms immune to data uncertainties. However, each discipline uses the term in a, slightly or totally, different context.

The purpose of this monograph is to summarize the applications of robust optimization in data mining. For this we present the most popular algorithms such as least squares, linear discriminant analysis, principal component analysis, and support vector machines along with their robust counterpart formulation. For the problems that have been proved to be tractable we describe their solutions.

Our goal is to provide a guide for junior researchers interested in pursuing theoretical research in data mining and robust optimization. For this we assume minimal familiarity of the reader with the context except of course for some basic linear algebra and calculus knowledge. This monograph has been developed so that each chapter can be studied independent of the others. For completion we include two appendices describing some basic mathematical concepts that are necessary for having complete understanding of the individual chapters. This monograph can be used not only as a guide for independent study but also as a supplementary material for a technically oriented graduate course in data mining.

Orlando, FL   Petros Xanthopoulos
Gainesville, FL   Panos M. Pardalos
Norman, OK   Theodore B. Trafalis

Acknowledgments

Panos M. Pardalos would like to acknowledge the Defense Threat Reduction Agency (DTRA) and the National Science Foundation (NSF) for the funding support of his research.

Theodore B. Trafalis would like to acknowledge National Science Foundation (NSF) and the U.S. Department of Defense, Army Research Office for the funding support of his research.

Contents

Chapter 1
Introduction

Abstract Data mining (DM), conceptually, is a very general term that encapsulates a large number of methods, algorithms, and technologies. The common denominator among all these is their ability to extract useful patterns and associations from data usually stored in large databases. Thus DM techniques aim to provide knowledge and interesting interpretation of, usually, vast amounts of data. This task is crucial, especially today, mainly because of the emerging needs and capabilities that technological progress creates. In this monograph we investigate some of the most well-known data mining algorithms from an optimization perspective and we study the application of robust optimization (RO) in them. This combination is essential in order to address the unavoidable problem of data uncertainty that arises in almost all realistic problems that involve data analysis. In this chapter we provide some historical perspectives of data mining and its foundations and at the same time we "touch" the concepts of robust optimization and discuss its differences compared to stochastic programming.

1.1 A Brief Overview

Before we state the mathematical problems of this monograph, we provide, for the sake of completion, a historical and methodological overview of data mining (DM). Historically DM was evolved, in its current form, during the last few decades from the interplay of classical statistics and artificial intelligence (AI). It is worth mentioning that through this evolution process DM developed strong bonds with computer science and optimization theory. In order to study modern concepts and trends of DM we first need to understand its foundations and its interconnections with the four aforementioned disciplines.

P. Xanthopoulos et al., *Robust Data Mining*, SpringerBriefs in Optimization,
DOI 10.1007/978-1-4419-9878-1_1,
© Petros Xanthopoulos, Panos M. Pardalos, Theodore B. Trafalis 2013

1.1.1 Artificial Intelligence

The perpetual need/desire of human to create artificial machines/algorithms able to learn, decide, and act as humans, gave birth to AI. Officially AI was born in 1956 in a conference held at Dartmouth College. The term itself was coined by J. McCarthy during that conference. The goals of AI stated at this first conference, even today, might be characterized as superficial from a pessimist perspective or as challenging from an optimistic perspective. By reading again the proceedings of this conference, we can see the rough expectations of the early AI community: "To proceed on the basis of the conjecture that every aspect of learning or any other feature of intelligence can be so precisely described that a machine can be made to simulate it" [37]. Despite the fact that even today understanding the basic underlying mechanisms of cognition and human intelligence remain an open problem for computational and clinical scientists, this founding conference of AI stimulated the scientific community and triggered the development of algorithms and methods that became the foundations of modern machine learning. For instance, bayesian methods were developed and further studied as part of AI research. Computer programming languages like LISP [36] and PROLOG [14] were also developed for serving AI purposes, and algorithms such as perceptron [47], backpropagation [15], and in general artificial neural networks (ANN) were invented for the same purpose.

1.1.2 Computer Science/Engineering

In literature DM is often classified as a branch of computer science (CS). Indeed a lot of DM research has been driven by CS society. In addition to this, there were several advances of CS that boosted DM research. Database modeling together with smart search algorithms made possible the indexing and processing of massive databases [1,44]. The advances, in software level, of database modeling and search algorithms were accompanied by a parallel development of semiconductor technologies and computer hardware engineering.

In fact there is a feedback relation between DM and computer engineering that drives the research in both areas. Computer engineering provides cheaper and larger storage and processing power. On the other hand these new capabilities pose new problems for DM society, often related to the processing of such amounts of data. These problems create new algorithms and new needs for processing power that is in turns addressed by computer engineering society. The progress in this area can be best described by the so-called Moore's "law" (named after Intel's cofounder G. E. Moore) that predicted that the number of transistors on a chip will double every 24 months [39]. The predictions of this simple rule have been accurate at least until today (Fig. 1.1).

Similar empirical "laws" have been stated for hard drive capacity and hard drive price. Hard drive capacity increases ten times every 5 years and the cost

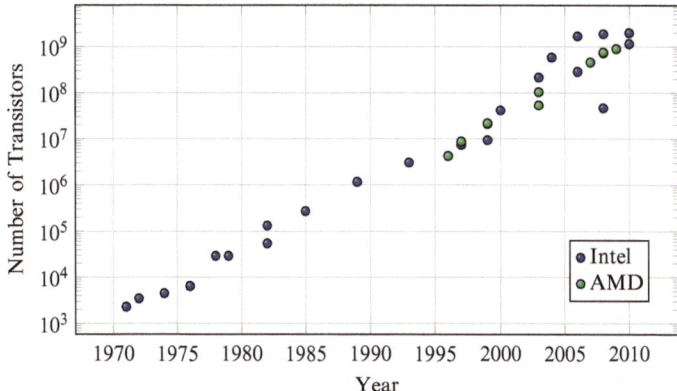

Fig. 1.1 Moore's "law" drives the semiconductor market even today. This plot shows the transistor count of several processors from 1970 until today for two major processor manufacturing companies (Intel and AMD). Data source: http://en.wikipedia.org/wiki/Transistor_count

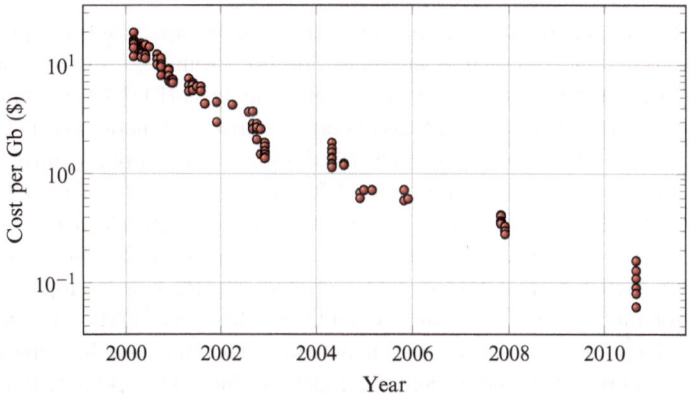

Fig. 1.2 Kryder's "law" describes the exponential decrease of computer storage cost over time. This rule is able to predict approximately the cost of storage space over the last decade

drops ten times every five years. This empirical observation is known as Kryder's "law" (Fig. 1.2) [61]. Similar rule which is related to network bandwidth per user (Nielsen's "law") indicates that it increases by 50% annually [40]. The fact that computer progress is characterized by all these exponential empirical rules is in fact indicative of the continuous and rapid transformation of DM's needs and capabilities.

1.1.3 Optimization

Mathematical theory of optimization is a branch of mathematics that was originally developed for serving the needs of operations research (OR). It is worth noting

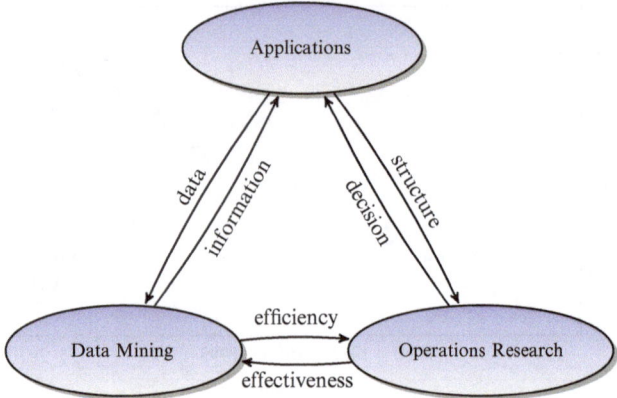

Fig. 1.3 The big picture. Scheme capturing the inderdependence among DM, OR, and the various application fields

that a large amount of data mining problems can be described as optimization problems, sometimes tractable, sometimes not. For example, principal component analysis (PCA) and Fisher's linear discriminant analysis (LDA) are formulated as minimization/ maximization problems of certain statistical functionals [11]. Support vector machines (SVMs) can be described as a convex optimization problem [60] and linear programming can be used for development of supervised learning algorithms [35]. In addition several optimization metaheuristics have been proposed for adjusting the parameters of supervised learning models [12]. On the other side, data mining methods are often used as preprocessing for before employing some optimization model (e.g., clustering). In addition a branch of DM involves network models and optimization problems on networks for understanding the complex relationships between the nodes and the edges. In this sense optimization is a tool that can be employed in order to solve DM problems. In a recent review paper the interplay of operations research data mining and applications was described by the scheme shown in Fig. 1.3 [41].

1.1.4 Statistics

Statistics set the foundation for many concepts broadly used in data mining. Historically, one of the first attempts to understand interconnection between data was Bayes analysis in 1763 [5]. Other concepts include regression analysis, hypothesis testing, PCA, and LDA. As discussed, in modern DM it is very common to maximize or minimize certain statistical quantities in order to achieve some clustering (grouping) or to find interconnections and patterns among groups of data.

1.2   A Brief History of Robustness

The term "robust" is used extensively in engineering and statistics literature. In engineering it is often used in order to denote error resilience in general, e.g., robust methods are these that are not affected much by small error interferences. In statistics robust is used to describe all these methods that are used when the model assumptions are not exactly true, e.g., variables follow exactly the assumed distribution (existence of outliers). In optimization (minimization of maximization) robustness is used in order to describe the problem of finding the best solution given that the problem data are not fixed but obtain their values within a well-defined uncertainty set. Thus if we consider the minimization problem (without loss of generality)

$$\min_{x \in \mathcal{X}} \ f(A,x) \tag{1.1a}$$

where A accounts for all the parameters of the problem that are considered to be fixed numbers, and $f(\cdot)$ is the objective function, the robust counterpart (RC) problem is going to be a min–max problem of the following form:

$$\min_{x \in \mathcal{X}} \max_{A \in \mathcal{A}} \ f(A,x) \tag{1.2a}$$

where $\mathcal{A}$ is the set of all admissible perturbations. The maximization problem over the parameters A corresponds, usually, to a worst case scenario. The objective of robust optimization is to determine the optimal solution when such a scenario occurs. In real data analysis problems it is very likely that data might be corrupted, perturbed, or subject to errors related to data acquisition. In fact most of the modern data acquisition methods are prone to errors. The most usual source of such errors is noise which is usually associated with the instrumentation itself or due to human factors (when the data collection is done manually). Spectroscopy, microarray technology, and electroencephalography (EEG) are some of the most commonly used data collection technologies that are subject to noise. Robust optimization is employed not only when we are dealing with data imprecisions but also when we want to provide stable solutions that can be used in case of input modification. In addition it can be used in order to avoid selection of "useless" optimal solutions i.e. solutions that change drastically for small changes of data. Especially in case where an optimal solution cannot be implemented precisely, due to technological constraints, we wish that the next best optimal solution will be feasible and very close to the one that is out of our implementation scope. For all these reasons, robust methods and solutions are highly desired.

In order to outline the main goal and idea of robust optimization we will use the well-studied example of linear programming (LP). In this problem we need to determine the global optimum of a linear function over the feasible region defined by a linear system.

$$\min \ c^{\mathrm{T}}x \tag{1.3a}$$

$$\text{s.t.} \ Ax = b \tag{1.3b}$$

$$x \geq 0 \tag{1.3c}$$

where $A \in \mathbb{R}^{n \times m}, b \in \mathbb{R}^n, c \in \mathbb{R}^m$. In this formulation x is the decision variable and A, b, c are the data and they have constant values. The LP for fixed data values can be solved efficiently by many algorithms (e.g., SIMPLEX) and has been shown that it can be solved in polynomial time [28].

In the case of uncertainty, we assume that data are not fixed but they can take any values within an uncertainty set with known boundaries. Then the robust counterpart (RC) problem is to find a vector x that minimizes (1.3a) for the "worst case" perturbation. This worst case problem can be stated as a maximization problem with respect to A, b, and c. The whole process can beformulated as the following min–max problem:

$$\min_{x} \max_{A,b,c} \ c^{\mathrm{T}}x \tag{1.4a}$$

$$\text{s.t.} \ Ax = b \tag{1.4b}$$

$$x \geq 0 \tag{1.4c}$$

$$A \in \mathscr{A}, b \in \mathscr{B}, c \in \mathscr{C} \tag{1.4d}$$

where $\mathscr{A}, \mathscr{B}, \mathscr{C}$ are the uncertainty sets of A, b, c correspondingly. problem (1.4) can be tractable or untractable based on the uncertainty sets properties. For example, it has been shown that if the columns of A follow ellipsoidal uncertainty constraints the problem is polynomially tractable [7]. Bertsimas and Sim showed that if the coefficients of A matrix are between a lower and an upper bound, then this problem can be still solved with linear programming [9]. Also Bertsimas et al. have shown that an uncertain LP with general norm bounded constraints is a convex programming problem [8]. For a complete overview of robust optimization, we refer the reader to [6]. In the literature there are numerous studies providing with theoretical or practical results on robust formulation of optimization problems. Among others mixed integer optimization [27], conic optimization [52], global optimization [59], linear programming with right-hand side uncertainty [38], graph partitioning [22], and critical node detection [21].

1.2.1 Robust Optimization vs Stochastic Programming

Here it is worth noting that robust optimization is not the only approach for handling uncertainty in optimization. In the robust framework the information about uncertainty is given in a rather deterministic form of worst case bounding constraints. In a different framework one might not require the solution to be feasible for all data realization but to obtain the best solution given that problem data are random variables following a specific distribution. This is of particular interest when the problem possesses some periodic properties and historical data are available. In this case the parameters of such a distribution could efficiently be estimated through some model fitting approach. Then a probabilistic description of the constraints can be obtained and the corresponding optimization problem can be classified as a stochastic programming problem. Thus the stochastic equivalent of the linear program (1.3a) will be:

$$\min_{x,t}\ t \tag{1.5a}$$

$$\text{s.t.}\ \Pr\{c^\mathsf{T}x \le t, Ax \le b\} \ge p \tag{1.5b}$$

$$x \ge 0 \tag{1.5c}$$

where c, A, and b are random variables that follow some known distribution, p is a nonnegative number less than 1 and $\Pr\{\cdot\}$ some legitimate probability function. This non-deterministic description of the problem does not guarrantee that the provided solution would be feasible for all data set realizations but provides a less conservative optimal solution taking into consideration the distribution-based uncertainties. Although the stochastic approach might be of more practical value in some cases, there are some assumptions made that one should be aware of [6]:

1. The problem must be of stochastic nature and that indeed there is a distribution hidden behind each variable.
2. Our solution depends on our ability to determine the correct distribution from the historic data.
3. We have to be sure that our problem accepts probabilistic solutions, i.e., a stochastic problem solution might not be immunized against a catastrophic scenario and a system might be vulnerable against rare event occurrence.

For this, the choice of the approach strictly depends on the nature of the problem as well as the available data. For an introduction to stochastic programming, we refer the reader to [10].

Chapter 2
Least Squares Problems

Abstract In this chapter we provide an overview of the original minimum least squares problem and its variations. We present their robust formulations as they have been proposed in the literature so far. We show the analytical solutions for each variation and we conclude the chapter with some numerical techniques for computing them efficiently.

2.1 Original Problem

In the original linear least squares (LLS) problem one needs to determine a linear model that approximates "best" a group of samples (data points). Each sample might correspond to a group of experimental parameters or measurements and each individual parameter to a feature or, in statistical terminology, to a predictor. In addition, each sample is characterized by an outcome which is defined by a real valued variable and might correspond to an experimental outcome. Ultimately we wish to determine a linear model able to issue outcome prediction for new samples. The quality of such a model can be determined by a minimum distance criterion between the samples and the linear model. Therefore if n data points, of dimension m each, are represented by a matrix $A \in \mathbb{R}^{n \times m}$ and the outcome variable by a vector $b \in \mathbb{R}^n$ (each entry corresponding to a row of matrix A), we need to determine a vector $x \in \mathbb{R}^m$ such that the residual error, expressed by some norm, is minimized. This can be stated as:

$$\min_{x} \|Ax - b\|_2^2 \tag{2.1}$$

where $\| \cdot \|_2$ is the Euclidean norm of a vector. The objective function value is also called residual and denoted $r(A, b, x)$ or just r. The geometric interpretation of this problem is to find a vector x such that the sum of the distances between the points represented by the rows of matrix A and the hyperplane defined by $x^{\mathrm{T}}w - b = 0$ (where w is the independent variable) is minimized. In this sense this problem is a

P. Xanthopoulos et al., *Robust Data Mining*, SpringerBriefs in Optimization,
DOI 10.1007/978-1-4419-9878-1_2,
© Petros Xanthopoulos, Panos M. Pardalos, Theodore B. Trafalis 2013

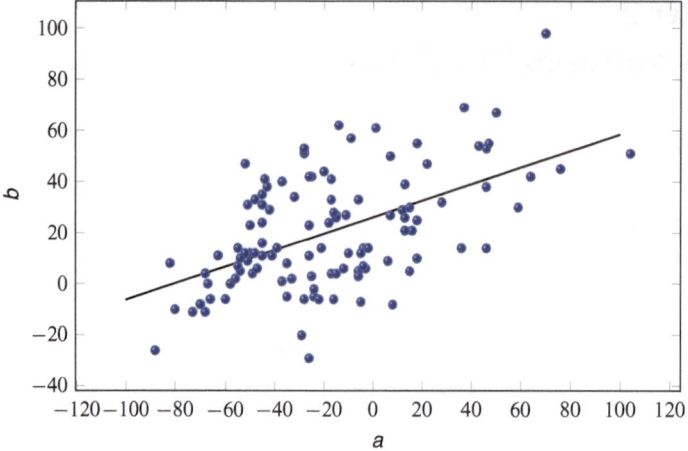

Fig. 2.1 The single input single outcome case. This is a 2D example the predictor represented by the *a* variable and the outcome by vertical axis *b*

first order polynomial fitting problem. Then by determining the optimal *x* vector will be able to issue predictions for new samples by just computing their inner product with *x*. An example in two dimensions (2D) can be seen in Fig. 2.1. In this case the data matrix will be $A = [a \; e] \in \mathbb{R}^{n \times 2}$ where *a* is the predictor variable and *e* a column vector of ones that accounts for the constant term.

The problem can be solved, in its general form, analytically since we know that the global minimum will be at a Karush–Kuhn–Tucker (KKT) point (since the problem is convex and unconstrained) the Lagrangian equation $\mathscr{L}_{\text{LLS}}(x)$ will be given by the objective function itself and the KKT points can be obtained by solving the following equation:

$$\frac{d\mathscr{L}_{\text{LLS}}(x)}{dx} = 0 \Leftrightarrow 2A^{\mathsf{T}}Ax = A^{\mathsf{T}}b \tag{2.2}$$

In case that *A* is of full row rank, that is $\text{rank}(A) = n$, matrix $A^{\mathsf{T}}A$ is invertible and we can write:

$$x_{\text{LLS}} = \left(A^{\mathsf{T}}A\right)^{-1}A^{\mathsf{T}}b \triangleq A^{\dagger}b \tag{2.3}$$

Matrix $A^{\dagger}$ is also called pseudoinverse or Moore–Penrose matrix. It is very common that the full rank assumption is not always valid. In such case the most common way to address the problem is through regularization. One of the most famous regularization techniques is the one known as Tikhonov regularization [55]. In this case instead of problem (2.1) we consider the following problem:

$$\min_{x} \left(\|Ax - b\|^2 + \delta\|x\|^2\right) \tag{2.4}$$

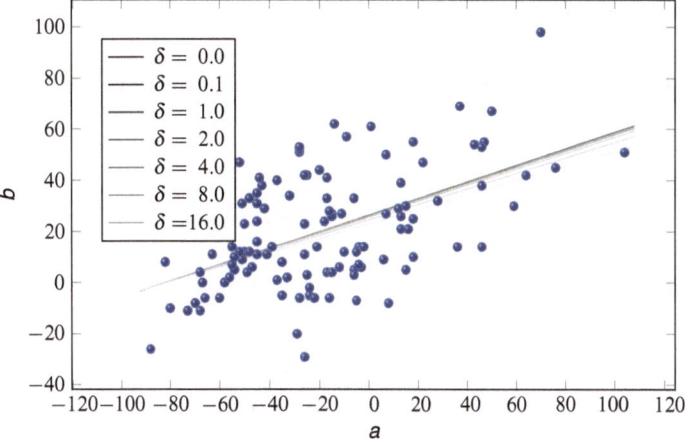

Fig. 2.2 LLS and regularization. Change of linear least squares solution with respect to different δ values. As we can observe, in this particular example, the solution hyperplane is slightly perturbed for different values of δ

by using the same methodology we obtain:

$$\frac{d\mathscr{L}_{RLLS}(x)}{dx} = 0 \Leftrightarrow A^T(Ax - b) + \delta Ix = 0 \Leftrightarrow (A^TA + \delta I)x = A^Tb \qquad (2.5)$$

where I is a unit matrix of appropriate dimension. Now even in case that A^TA is not invertible we can compute x by

$$x_{RLLS} = (A^TA + \delta I)^{-1}A^Tb \qquad (2.6)$$

This type of least square solution is also known as ridge regression. The parameter δ controls the trade-off between optimality and stability. Originally regularization was proposed in order to overcome this practical difficulty that arises in real problems and it is related to rank deficiency described earlier. The value of δ is determined usually by trial and error and its magnitude is smaller compared to the entries of data matrix. In Fig. 2.2 we can see how the least squares plane changes for different values of delta.

In Sect. 2.5 we will examine the relation between robust linear least squares and robust optimization.

2.2 Weighted Linear Least Squares

A slight, and more general, modification of the original least squares problem is the weighted linear least squares problem (WLLS). In this case we have the following minimization problem:

$$\min_x r^T W r = \min_x (Ax - b)^T W (Ax - b) = \min_x \|W^{1/2}(Ax - b)\| \qquad (2.7)$$

where W is the weight matrix. Note that this is a more general formulation since for $W = I$ the problem reduces to (2.1). The minimum can be again obtained by the solution of the corresponding KKT systems which is:

$$2A^T W (Ax - b) = 0 \qquad (2.8)$$

and gives the following solution:

$$x_{\text{WLLS}} = (A^T W A)^{-1} A^T W b \qquad (2.9)$$

assuming that $A^T W A$ is invertible. If this is not the case regularization is employed resulting in the following regularized weighted linear least squares (RWLLS) problem

$$\min_x \left(\|W^{1/2}(Ax - b)\|^2 + \delta \|x\|^2 \right) \qquad (2.10)$$

that attains its global minimum for

$$x_{\text{RWLLS}} = (A^T W A + \delta I)^{-1} A W b \qquad (2.11)$$

Next we will discuss some practical approaches for computing least square solution for all the discussed variations of the problem.

2.3 Computational Aspects of Linear Least Squares

Least squares solution can be obtained by computing an inverse matrix and applying a couple of matrix multiplications. However, in practice, direct matrix inversion is avoided, especially due to the high computational cost and solution instabilities. Here we will describe three of the most popular methods used for solving the least squares problems.

2.3.1 Cholesky Factorization

When matrix A is of full rank, then AA^T is invertible and can be decomposed through Cholesky decomposition in a product LL^T where L is a lower triangular matrix. Then (2.2) can be written as:

$$LL^Tx = A^Tb \tag{2.12}$$

that can be solved by a forward substitution followed by a backward substitution. In case that A is not of full rank, then this procedure can be applied to the regularized problem (2.5).

2.3.2 QR Factorization

An alternative method is the one of QR decomposition. In this case we decompose matrix AA^T into a product of two matrices where the first matrix Q is orthogonal and the second matrix R is upper triangular. This decomposition again requires data matrix A to be of full row rank. Orthogonal matrix Q has the property $QQ^T = I$ thus the problem is equivalent to

$$Rx = Q^TA^Tb \tag{2.13}$$

and it can be solved by backward substitution.

2.3.3 Singular Value Decomposition

This last method does not require full rank of matrix A. It uses the singular value decomposition of A:

$$A = U\Sigma V^T \tag{2.14}$$

where U and V are orthogonal matrices and Σ is diagonal matrix that has the singular values. Every matrix with real elements has an SVD and furthermore it can be proved that a matrix is of full row rank if and only if all of its singular values are nonzero. Substituting with its SVD decomposition we get:

$$AA^Tx = (U\Sigma V^T)(V\Sigma U^T)x = U\Sigma^2 U^Tx = A^Tb \tag{2.15}$$

and finally

$$x = U(\Sigma^2)^\dagger U^TA^Tb \tag{2.16}$$

The matrix $(\Sigma^2)^\dagger$ can be computed easily by inverting its nonzero entries. If A is of full rank then all singular values are non-zero and $(\Sigma^2)^\dagger = (\Sigma^2)^{-1}$. Although SVD can be applied to any kind of matrix it is computationally expensive and sometimes is not preferred especially when processing massive datasets.

2.4   Least Absolute Shrinkage and Selection Operator

An alternative regularization technique for the same problem is the one of least absolute shrinkage and selection operator (LASSO) [54]. In this case the regularization term contains a first norm term $\delta\|x\|_1$. Thus we have the following minimization problem:

$$\min_x \left(\|Ax - b\|^2 + \delta\|x\|_1 \right) \tag{2.17}$$

Although this problem cannot be solved analytically as the one obtained after Tikhonov regularization, sometimes it is preferred as it provides sparse solutions. That is the solution x vector obtained by LASSO has more zero entries. This approach has a lot of applications in compressive sensing [2, 16, 34]. As it will be discussed later this regularization possesses further robust properties as it can be obtained through robust optimization for a specific type of data perturbations.

2.5   Robust Least Squares

2.5.1   Coupled Uncertainty

Now we will study the robust version of problem (2.1). The results presented here were first described in [20] and similar results were independently obtained in [18]. At the end we describe some extensions that were first described in [17]. As we discussed earlier the RC formulation of a problem involves solution of a worst case scenario problem. This is expressed by a min–max (or max–min) type problem where the outer min (max) problem refers to the original one whereas the inner max (min) to the worst admissible scenario. For the least squares case the generic RC formulation can be described from the following problem:

$$\min_x \max_{\Delta A \in \mathscr{U}_A, \Delta b \in \mathscr{U}_b} \|(A + \Delta A)x - (b + \Delta b)\|_2 \tag{2.18}$$

where $\Delta A, \Delta B$ are perturbation matrices and $\mathscr{U}_A. \mathscr{U}_B$ are sets of admissible perturbations. As in many robust optimization problems, the structural properties of $\mathscr{U}_A, \mathscr{U}_B$ are important for the computational tractability of the problem. Here we study the case where the two perturbation matrices are unknown but their norm is bounded by a known constant. Thus we have the following optimization problem:

$$\min_x \max_{\|\Delta A\| \leq \rho_A, \|\Delta b\| \leq \rho_b} \|(A + \Delta A)x - (b + \Delta b)\|_2 \tag{2.19}$$

This type of uncertainty is often called coupled uncertainty because the uncertainty information is not given in terms of each sample individually but in terms of the whole data matrix. This can be interpreted as having a total uncertainty "budget" which not required to be distributed evenly among the dataset. Under this assumption we do not have any particular information for individual data points and the resulting solution to this problem can be extremely conservative. First we will reduce problem (2.19) to a minimization problem through the following lemma

Lemma 2.1. *The problem* (2.19) *is equivalent to the following:*

$$\min_{x} \left(\|Ax - b\| + \rho_A \|x\| + \rho_b \right) \tag{2.20}$$

Proof. From triangular inequality we can obtain an upper bound on the objective function of (2.19):

$$\|(A + \Delta A)x - (b + \Delta b)\| \leq \|Ax - b\| + \|\Delta Ax - \Delta b\| \tag{2.21}$$

$$\leq \|Ax - b\| + \|\Delta A\| \|x\| + \|\Delta b\| \tag{2.22}$$

$$\leq \|Ax - b\| + \rho_A \|x\| + \rho_B \tag{2.23}$$

Now if in the original problem (2.19) we set

$$\Delta A = \frac{Ax - b}{\|Ax - b\|} \frac{x^{\mathrm{T}}}{\|x\|} \rho_A, \quad \Delta b = -\frac{Ax - b}{\|Ax - b\|} \rho_B \tag{2.24}$$

we get

$$\|(A + \Delta A)x - (b + \Delta b)\| = \|Ax - b + \Delta Ax - \Delta b\|$$

$$= \|Ax - b\| \left(1 + \frac{\|x\|}{\|Ax - b\|} \rho_A + \frac{1}{\|Ax - b\|} \rho_B \right)$$

$$= \|Ax - b\| + \rho_A \|x\| + \rho_B \tag{2.25}$$

This means that the upper bound obtained by the triangular inequality can be achieved by (2.24). Since the problem is convex, this will be its global optimum.

We can easily observe that the point (2.24) satisfies the optimality conditions. Since problem (2.20) is unconstrained, its Lagrangian will be the same as the cost function. Since this function is convex we just need to examine the points for which the derivative is equal to zero and consider separate cases for the non-differentiable points. At the points where the cost function is differentiable we have:

$$\frac{\partial \mathscr{L}_{\mathrm{RLLS}}(x)}{\partial x} = 0 \Leftrightarrow \frac{A^{\mathrm{T}}(Ax - b)}{\|Ax - b\|} + \frac{x}{\|x\|} \rho_A = 0 \tag{2.26}$$

From this last expression we require $x \neq 0$ and $Ax \neq b$ (we will deal with this cases later). If we solve with respect to x, we obtain:

$$\frac{1}{\|Ax - b\|} \left(A^{\mathrm{T}}(Ax - b) + x \frac{\|Ax - b\|}{\|x\|} \rho_A \right) = 0 \tag{2.27}$$

or

$$\left(A^{\mathrm{T}}A + \rho_A \frac{\|Ax - b\|}{\|x\|}I\right)x = A^{\mathrm{T}}b \tag{2.28}$$

and finally

$$x = (A^{\mathrm{T}}A + \mu I)^{-1}A^{\mathrm{T}}b, \quad \text{where} \quad \mu = \frac{\|Ax - b\|}{\|x\|}\rho_A \tag{2.29}$$

In case that $Ax = b$ the solution is given by $x = A^{\dagger}b$ where $A^{\dagger}$ is the Moore–Penrose or pseudoinverse matrix of A. Therefore we can summarize this result in the following lemma:

Lemma 2.2. *The optimal solution to problem* (2.20) *is given by:*

$$x = \begin{cases} A^{\dagger}b & \textit{if } Ax = b \\ (A^{\mathrm{T}}A + \mu I)^{-1}A^{\mathrm{T}}b, & \mu = \rho_A \frac{\|Ax - b\|}{\|x\|} \; \textit{otherwise} \end{cases} \tag{2.30}$$

Since in this last expression μ is a function of x we need to provide a way in order to tune it. For this we need to use the singular value decomposition of data matrix A:

$$A = U\begin{bmatrix} \Sigma \\ 0 \end{bmatrix}V^{\mathrm{T}} \tag{2.31}$$

where Σ is the diagonal matrix that contains the singular values of A in descending order. In addition we partition the vector $U^{\mathrm{T}}b$ as follows:

$$\begin{bmatrix} b_1 \\ b_2 \end{bmatrix} = U^{\mathrm{T}}b \tag{2.32}$$

where b_1 contains the first n elements and b_2 the rest $m - n$. Now using this decompositions we will obtain two expressions for the numerator and the denominator of μ. First for the denominator:

$$x = (A^{\mathrm{T}}A + \mu I)^{-1}A^{\mathrm{T}}b = \left(V\Sigma^2 V^{\mathrm{T}} + \mu I\right)^{-1}V\Sigma b_1 = V\left(\Sigma^2 + \mu I\right)^{-1}\Sigma b_1 \tag{2.33}$$

the norm will be given from

$$\|x\| = \|\Sigma(\Sigma^2 + \mu I)^{-1}\| \tag{2.34}$$

and for the numerator

$$Ax - b = U\begin{bmatrix} \Sigma \\ 0 \end{bmatrix}V^{\mathrm{T}}V\left(\Sigma^2 + \mu I\right)^{-1}\Sigma b_1 - b \tag{2.35}$$

$$= U\left(\begin{bmatrix} \Sigma \\ 0 \end{bmatrix}\left(\Sigma^2 + \mu I\right)^{-1}\Sigma b_1 - U^{\mathrm{T}}b\right) \tag{2.36}$$

$$= U\left(\begin{bmatrix} \Sigma(\Sigma^2 + \mu I)^{-1}\Sigma b_1 - b_1 \\ -b_2 \end{bmatrix}\right) \tag{2.37}$$

$$= U\begin{bmatrix} -\mu(\Sigma^2 + \mu I)^{-1}b_1 \\ -b_2 \end{bmatrix} \tag{2.38}$$

and for the norm

$$\|Ax - b\| = \sqrt{\|b_2\|^2 + \alpha^2\|(\Sigma^2 + \mu I)^{-1}b_1\|^2} \tag{2.39}$$

Thus μ will be given by:

$$\mu = \frac{\|Ax - b\|}{\|x\|} = \rho_A \frac{\sqrt{\|b_2\|^2 + \alpha^2\|(\Sigma^2 + \mu I)^{-1}b_1\|^2}}{\|\Sigma(\Sigma^2 + \mu I)^{-1}b_1\|} \tag{2.40}$$

Note that in the present analysis we assume that data matrix A is of full rank. If this is not the case similar analysis can be performed (for details, see [17]). The final solution can be obtained by the solution of (2.40) computationally. Next we will present some variations of the original least squares problem that are discussed in [17].

2.6   Variations of the Original Problem

In [17] authors introduced least square formulation for slightly different perturbation scenarios. For example, in the case of the weighted least squares problem with weight uncertainty one is interested to find:

$$\min_{x} \max_{\|\Delta W\| \leq \rho_W} \|((W + \Delta W)(Ax - b))\| \tag{2.41}$$

using the triangular inequality we can obtain an upper bound:

$$\|(W + \Delta W)(Ax - b)\| \leq \|W(Ax - b)\| + \|\Delta W(Ax - b)\| \tag{2.42}$$

$$\leq \|W(Ax - b)\| + \rho_W\|Ax - b\| \tag{2.43}$$

Thus the inner maximization problem reduces to the following problem:

$$\min_{x} (\|W(Ax - b)\| + \rho_W\|Ax - b\|) \tag{2.44}$$

by taking the corresponding KKT conditions, similar to previous analysis, we obtain:

$$\frac{\partial \mathcal{L}_{\mathrm{WLLS}}(x)}{\partial x} = \frac{\partial \|W(Ax-b)\|}{\partial x} + \frac{\partial \|Ax-b\|}{\partial x} \tag{2.45}$$

$$= \frac{A^{\mathrm{T}}W^{\mathrm{T}}(WAx-Wb)}{\|W(Ax-b)\|} + \rho_W \frac{A^{\mathrm{T}}(Ax-b)}{\|Ax-b\|} \tag{2.46}$$

By solving the equation

$$\frac{\partial \mathcal{L}_{\mathrm{WLLS}}(x)}{\partial x} = 0 \tag{2.47}$$

we find that the solution should satisfy

$$A^{\mathrm{T}}(W^{\mathrm{T}}W+\mu I)Ax = A^{\mathrm{T}}(W^{\mathrm{T}}W+\mu I)b \ \text{ where } \ \mu_w = \frac{\|W(Ax-b)\|}{\|Ax-b\|} \tag{2.48}$$

Giving the expression for x

$$x = \begin{cases} A^{\dagger}b & \text{if } Ax=b \\ (WA)^{\dagger}Wb & \text{if } WAx=Wb \\ \left(A^{\mathrm{T}}(W^{\mathrm{T}}W+\mu_w I)A\right)^{-1}A^{\mathrm{T}}\left(W^{\mathrm{T}}W+\mu_w I\right)b & \text{otherwise} \end{cases} \tag{2.49}$$

where μ_w is defined in (2.48). The solution for the last one can be obtained through similar way as for the original least squares problem. In another variation of the problem the uncertainty can be given with respect to matrix A but in multiplicative form. Thus the robust optimization problem for this variation can be stated as follows:

$$\min_{x} \max_{\|\Delta A\| \le \rho_A} \|(I+\Delta A)Ax-b\| \tag{2.50}$$

which can be reduced to the following minimization problem:

$$\min_{x} (\|Ax-b\| + \rho_A\|Ax\|) \tag{2.51}$$

then by similar analysis we obtain:

$$\frac{\partial \mathcal{L}_{\mathrm{MLLS}}(x)}{\partial x} = \frac{A^{\mathrm{T}}(Ax-b)}{\|A^{\mathrm{T}}(Ax-b)\|} + \rho_A \frac{A^{\mathrm{T}}Ax}{\|Ax\|} = 0 \tag{2.52}$$

and finally

$$x = \begin{cases} (A^{\mathrm{T}}A)^{\dagger}b & \text{if } A^{\mathrm{T}}Ax=A^{\mathrm{T}}b \\ \left(A^{\mathrm{T}}A(1+\mu_A)\right)^{-1}A^{\mathrm{T}}b, \mu_A = \frac{\|A^{\mathrm{T}}(Ax-b)\|}{\|Ax\|} & \text{otherwise} \end{cases} \tag{2.53}$$

2.6.1 Uncoupled Uncertainty

In the case that we have specific knowledge for the uncertainty bound of each data point separately we can consider the corresponding problem. The solution for this type of uncertainty reveals a very interesting connection between robustness and LASSO regression. Originally this result was obtained by Xu et al. [63]. Let us consider the least squares problem where the uncoupled uncertainties exist only with respect to the rows of the data matrix A:

$$\min_x \max_{\Delta A \in \mathscr{A}} ||(A + \Delta A)x - b||_2 \tag{2.54}$$

where the uncertainty set $\mathscr{A}$ is defined by:

$$\mathscr{A} \triangleq \{(\delta_1, \delta_2, \dots, \delta_m) | ||\delta_i|| \le \rho_i\} \tag{2.55}$$

For the maximization problem and for a fixed vector x:

$$\max_{\Delta A \in \mathscr{A}} ||(A + \Delta A)x - b||_2 = \max_{\Delta A \in \mathscr{A}} ||Ax - b + \Delta x||_2 \tag{2.56}$$

$$= \max_{\Delta A \in \mathscr{A}} ||Ax - b + \sum_{i=1}^m x_i \delta_i||_2 \tag{2.57}$$

$$\le \max_{\Delta A \in \mathscr{A}} ||Ax - b||_2 + \sum_{i=1}^m ||x_i \delta_i||_2 \tag{2.58}$$

$$\le \max_{\Delta A \in \mathscr{A}} ||Ax - b||_2 + \sum_{i=1}^m |x_i| \cdot \rho_i \tag{2.59}$$

This provides an upper bound for the objective function. This bound is obtained by proper use of the triangular inequality. On the other side if we let

$$u = \begin{cases} \frac{Ax - b}{||Ax - b||_2} & \text{if } Ax \ne b \\ \text{any unit norm vector otherwise} \end{cases} \tag{2.60}$$

Next we define the perturbation being equal to

$$\delta_i^* \triangleq \begin{cases} -c_i \cdot \text{sign}(x)u & \text{if } x_i \ne 0 \\ -c_i u & o/w \end{cases} \tag{2.61}$$

This perturbation belongs to the set of admissible perturbations since $||\delta_i^*||_2 = c_i$. If we set the perturbation in the maximization problem of (2.54) equal to (2.61), we get:

$$\max_{\Delta A \in \mathscr{A}} \|(A + \Delta A)x - b\|_2 \geq \|(A + \Delta A)\hat{x} - b\|_2 \tag{2.62}$$

$$= \|(A + (\delta_1^*, \delta_2^*, \ldots, \delta_m^*))x - b\|_2 \tag{2.63}$$

$$= \|Ax - b + \sum_{i:x_i \neq 0} (-x_i \cdot \mathrm{sgn}(x_i)u)\|_2 \tag{2.64}$$

$$= \|Ax - b + u \cdot \sum_{i=1}^{m} c_i|x_i|\|_2 \tag{2.65}$$

$$= \|Ax - b\|_2 + \sum_{i=1}^{m} c_i|x_i| \tag{2.66}$$

The last equation combined with (2.59) yields that the maximization problem:

$$\max_{\Delta A \in \mathscr{A}} \|(A + \Delta A)x - b\|_2 \tag{2.67}$$

attains its maximum for the point $\Delta A = (\delta_1, \delta_2, \ldots, \delta_m)$ where $\delta_i, i = 1, \ldots, m$ is defined by (2.61). This proves that the original problem can be written as:

$$\min_x \max_{\Delta A \in \mathscr{A}} \|(A + \Delta A)x - b\|_2 = \min_x \left\{ \|Ax - b\|_2 + \sum_{i=1}^{m} c_i \cdot |x_i| \right\} \tag{2.68}$$

The last is nothing but a regularized linear least squares problem with l_1 regularization term. The last relation not only proves another interesting connection between regularization and robustness but also suggests a practical method for adjusting the regularization parameter in case that we have prior knowledge of data uncertainty.

As pointed out by the authors in [63] the above result can be generalized for any arbitrary norm. Thus the robust regression problem

$$\min_x \max_{\Delta A \in \mathscr{U}_p} \|(A + \Delta A)x - b\|_p \tag{2.69}$$

with

$$\mathscr{U}_p \triangleq \{(\delta_1, \delta_2, \ldots, \delta_m) | \|\delta_i\|_p \leq \rho_i\} \tag{2.70}$$

is equivalent to the following problem

$$\min_x \left\{ \|Ax - b\|_p + \sum_{i=1}^{m} c_i \cdot |x_i| \right\} \tag{2.71}$$

This shows that LASSO type regularization can be the robust equivalent of a general regression problem regardless the norm given that the induced perturbations are defined as in (2.70).

Chapter 3
Principal Component Analysis

Abstract The principal component analysis (PCA) transformation is a very common and well-studied data analysis technique that aims to identify some linear trends and simple patterns in a group of samples. It has application in several areas of engineering. It is popular from computational perspective as it requires only an eigendecomposition or singular value decomposition. There are two alternative optimization approaches for obtaining principal component analysis solution, the one of variance maximization and the one of minimum error formulation. Both start with a "different" initial objective and end up providing the same solution. It is necessary to study and understand both of these alternative approaches. In the second part of this chapter we present the robust counterpart formulation of PCA and demonstrate how such a formulation can be used in practice in order to produce sparse solutions.

3.1 Problem Formulations

In this section we will present the two alternative formulation for the principal component analysis (PCA). Both of them are based on different optimization criteria, namely maximum variance and minimum error, but the final solution is the same. The PCA transformation was originally proposed by Pearson in 1901 [43], and it is still used until today in its generic form or as a basis for more complicated data mining algorithmic scheme. It offers a very basic interpretation of data allowing to capture simple linear trends (Fig. 3.1). At this point we need to note that we assume that the mean of the data samples is equal to zero. In case this is not true we need to subtract the sample mean as part of preprocessing.

P. Xanthopoulos et al., *Robust Data Mining*, SpringerBriefs in Optimization,
DOI 10.1007/978-1-4419-9878-1_3,
© Petros Xanthopoulos, Panos M. Pardalos, Theodore B. Trafalis 2013

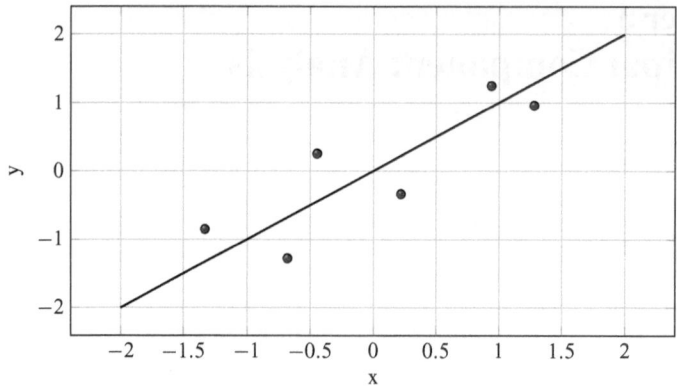

Fig. 3.1 Two paths for PCA. PCA has two alternative optimization formulations that result in the same outcome. One is to find a space where the projection of the original data will have maximum variance and the second is to find the subspace such that the projection error is minimized

3.1.1   Maximum Variance Approach

In this case we try to find a subspace of dimensionality $p < m$ for which the variability of the projection of the points is maximized. If we denote with $\bar{x}$ the sample mean:

$$\bar{x} = \frac{1}{n} \sum_{i=1}^{n} x_i \tag{3.1}$$

then the variance of the projected data on the subspace defined by the direction vector u will be:

$$\frac{1}{n} \sum_{i=1}^{n} \left(u^T x_i - u^T \bar{x} \right)^2 = \frac{1}{n} \sum_{i=1}^{n} \left(u^T (x_i - \bar{x}) \right)^2 = u^T \left(\frac{\sum_{i=1}^{n} (x_i - \bar{x})^T (x_i - \bar{x})}{n} \right) u \tag{3.2}$$

and given that the variance covariance matrix is defined by:

$$S = \frac{\sum_{i=1}^{n} (x_i - \bar{x})^T (x_i - \bar{x})}{n} \tag{3.3}$$

Equation (3.2) can be written in matrix notation as:

$$u^T S u \tag{3.4}$$

If we restrict, without loss of generality, our solution space just to the vectors u with Euclidean unit norm, then PCA problem can be expressed as the following optimization problem:

$$\max_{u}\ u^{\mathrm{T}}Su \tag{3.5a}$$

$$\text{s.t. } u^{\mathrm{T}}u = 1 \tag{3.5b}$$

The Lagrangian $\mathscr{L}_{\mathrm{PCA}}(u,\lambda)$ for this problem will be:

$$\mathscr{L}_{\mathrm{PCA}}(u,\lambda) = u^{\mathrm{T}}Su + \lambda(u^{\mathrm{T}}u - 1) = 0 \tag{3.6}$$

where λ is the Lagrange multiplier associated with the single constraint of the problem. The optimal points will be given by the roots of the Lagrangian (since S is positive semidefinite the problem is convex minimization). Thus

$$Su = \lambda u \tag{3.7}$$

This equation is satisfied by all the eigenpairs $(\lambda_n, u_n), i = 1,\ldots,n$ where

$$\lambda_1 \le \lambda_2 \le \cdots \le \lambda_n \tag{3.8}$$

are the ordered eigenvalues and u_i's are the corresponding eigenvectors. The objective function is maximized for $u = u_n, \lambda = \lambda_n$ and the optimal objective function value is $u_n^{\mathrm{T}}Su_n = u_n^{\mathrm{T}}\lambda_n u_n = \lambda_n\|u_n\|^2 = \lambda_n$.

3.1.2 Minimum Error Approach

An alternative derivation of PCA can be achieved through a different path. In this approach the objective is to rotate the original axis system such that the projection error of the dataset to the rotated system will be minimized. Thus we define a set of basis vector $\{u_i\}_{i=1}^m$. As soon as we do this we are able to express every point, including our dataset points, as a linear combination of the basis vectors.

$$x_k = \sum_{i=1}^m a_{ki}u_i = \sum_{i=1}^m \left(x_k^{\mathrm{T}}u_i\right)u_i, \quad k = 1,\ldots,n \tag{3.9}$$

Our purpose is to approximate every point x_k with $\tilde{x}_k$ using just a subset $p < m$ of the basis. Thus the approximation will be:

$$\tilde{x}_k = \sum_{i=1}^p \left(x_k^{\mathrm{T}}u_i\right)u_i + \sum_{i=p+1}^m \left(\bar{x}^{\mathrm{T}}u_i\right)u_i, \quad k = 1,\ldots,n \tag{3.10}$$

where $\bar{x}$ is the sample mean. The approximation error can be computed through a squared Euclidean norm summation over all data points:

$$\sum_{k=1}^n \|x_k - \tilde{x}_k\|^2. \tag{3.11}$$

We can obtain a more compact expression for $x_k - \tilde{x}_k$:

$$x_k - \tilde{x}_k = \sum_{i=1}^{m} \left(x_k^{\mathrm{T}} u_i \right) u_i - \left(\sum_{i=1}^{p} \left(x_n^{\mathrm{T}} u_i \right) u_i + \sum_{i=p+1}^{m} \left(\bar{x}^{\mathrm{T}} u_i \right) u_i \right) \tag{3.12a}$$

$$= \sum_{i=p+1}^{m} \left((x_k^{\mathrm{T}} u_i) u_i - (\bar{x}^{\mathrm{T}} u_i) u_i \right) \tag{3.12b}$$

$$= \sum_{i=p+1}^{m} \left((x_k - \bar{x})^{\mathrm{T}} u_i \right) u_i. \tag{3.12c}$$

Then the solution can be estimated by minimizing (3.11) and by constraining the solution to the vectors of unit Euclidean norm

$$\min_u \sum_{i=p+1}^{m} u_i^{\mathrm{T}} S u_i = \max_u \sum_{i=1}^{p} u_i^{\mathrm{T}} S u_i \tag{3.13a}$$

$$\text{s.t. } u^{\mathrm{T}} u = 1 \tag{3.13b}$$

This optimization problem is similar to the one obtained through the maximum variance approach (the only difference is that we are looking for the first p components instead of just one) and the solution is given by the first p eigenvectors that correspond to the p highest eigenvalues (can be proved through analytical solution of KKT system).

3.2   Robust Principal Component Analysis

Now we will describe a robust optimization approach for PCA transformation. Again we need to clarify that the purpose of this work is to investigate the application of robust optimization in the PCA transformation. There have been several robust PCA papers in the literature that deal with the application of robust statistics in PCA [26] and they are of interest when outliers are present in the data. Unlike supervised learning approaches like SVM, where the objective is to find the optimal solution for the worst case scenario, the purpose of robust formulation of PCA, as described in [3], is to provide components explaining data variance while at the same time are as sparse as possible. This is in general called sparse principal component analysis (SPCA) transformation. By sparse solutions we mean the vectors with large number of zeros. In general sparsity can be enforced through different methods. Sparsity is a desired property, especially in telecommunications, because it allows more efficient compression and faster data transmission. An SPCA formulation can be obtained if we add a cardinality constraint that strictly enforces sparsity. That is:

$$\max\ u^T S u \tag{3.14a}$$

$$\text{s.t}\ u^T u = 1 \tag{3.14b}$$

$$\text{card}(x) \leq k \tag{3.14c}$$

where $\text{card}(\cdot)$ is the cardinality function and k is parameter defining the maximum allowed component cardinality. Matrix S is the covariance matrix defined previously and u is the decision variable vector. Alternatively this problem can be casted as a semidefinite programming problem as follows:

$$\max\ \text{Tr}(US) \tag{3.15a}$$

$$\text{s.t}\ \text{Tr}(U) = 1 \tag{3.15b}$$

$$\text{card}(U) \leq k^2, \tag{3.15c}$$

$$U \succeq 0, \text{Rank}(X) = 1 \tag{3.15d}$$

where U is the decision variable matrix and $\succeq$ denotes that the matrix is positive semidefinite (i.e. $a^T X a \geq 0, \forall a \in \mathbb{R}^n$). Indeed the solution to the original problem can be obtained from the second one since conditions (3.15d) guarantee that $U = u \cdot u^T$. Instead of strictly constraining the cardinality we will demand $e^T \cdot \text{abs}(U) \cdot e \leq k$ (where e is the vector of 1's and $\text{abs}(\cdot)$ returns the matrix whose elements are the absolute values of the original matrix). In addition we will drop the rank constraint as this is also a tough to handle constraint. We obtain the following relaxation of the original problem:

$$\max\ \text{Tr}(US) \tag{3.16a}$$

$$\text{s.t}\ \text{Tr}(U) = 1 \tag{3.16b}$$

$$e^T \cdot \text{abs}(U) \cdot e \leq k \tag{3.16c}$$

$$U \succeq 0 \tag{3.16d}$$

the last relaxed problem is a semidefinite program with respect to matrix variable U. We can rewrite it as follows:

$$\max\ \text{Tr}(US) \tag{3.17a}$$

$$\text{s.t}\ \text{Tr}(U) = 1 \tag{3.17b}$$

$$e^T \cdot \text{abs}(U) \cdot e \leq k \tag{3.17c}$$

$$U \succeq 0 \tag{3.17d}$$

If we now remove the constraint (3.17c) and add a penalization term to the objective function, we obtain the following relaxation:

$$\max \; \mathrm{Tr}(US) - \rho e^{\mathrm{T}} \cdot \mathrm{abs}(U) \cdot e \qquad (3.18a)$$

$$\text{s.t } \mathrm{Tr}(U) = 1 \qquad (3.18b)$$

$$U \succeq 0 \qquad (3.18c)$$

where ρ is parameter (Lagrange multiplier) determining the penalty's magnitude. By taking the dual of this problem we can have a better understanding for the nature of the problem.

$$\min \; \lambda^{\max}(S+V) \qquad (3.19a)$$

$$\text{s.t. } |V_{ij}| \leq \rho, \;\; i,j = 1,\dots,n \qquad (3.19b)$$

where $\lambda^{\max}(X)$ is the maximum eigenvalue of matrix X. The problem (3.18) can be rewritten as the following min–max problem:

$$\max_{X \succeq 0, \mathrm{Tr}(U)=1} \; \min_{|V_{ij}| \leq \rho} \; \mathrm{Tr}(U(S+V)) \qquad (3.20a)$$

More precisely the goal is to determine the component that corresponds to the maximum possible variance (which is the original PCA objective) by choosing the most sparse solution (according to the sparsity constraints).

Chapter 4
Linear Discriminant Analysis

Abstract In this chapter we discuss another popular data mining algorithm that
can be used for supervised or unsupervised learning. Linear Discriminant Analysis
(LDA) was proposed by R. Fischer in 1936. It consists in finding the projection
hyperplane that minimizes the interclass variance and maximizes the distance
between the projected means of the classes. Similarly to PCA, these two objectives
can be solved by solving an eigenvalue problem with the corresponding eigenvector
defining the hyperplane of interest. This hyperplane can be used for classification,
dimensionality reduction and for interpretation of the importance of the given
features. In the first part of the chapter we discuss the generic formulation of LDA
whereas in the second we present the robust counterpart scheme originally proposed
by Kim and Boyd. We also discuss the non linear extension of LDA through the
kernel transformation.

4.1 Original Problem

The linear discriminant analysis (LDA) is a fundamental data analysis method
originally proposed by R. Fisher for discriminating between different types of
flowers [23]. The intuition behind the method is to determine a subspace of lower
dimension, compared to the original data sample dimension, in which the data points
of the original problem are "separable" (Fig. 4.1). Separability is defined in terms
of statistical measures of mean value and variance. One of the advantages of LDA
is that the solution can be obtained by solving a generalized eigenvalue system.
This allows for fast and massive processing of data samples. In addition LDA can
be extended to non-LDA through the kernel trick [4]. The original algorithm was
proposed for binary class problems but multi-class generalizations have also been
proposed [45]. Here we will discuss both starting from the simple two-class case.

P. Xanthopoulos et al., *Robust Data Mining*, SpringerBriefs in Optimization,
DOI 10.1007/978-1-4419-9878-1_4,
© Petros Xanthopoulos, Panos M. Pardalos, Theodore B. Trafalis 2013

Fig. 4.1 The intuition behind LDA. Data samples in two dimensions are projected in a lower dimension space (*line*). The *line* has to be chosen so that the projection maximizes the "separability" of the projected samples

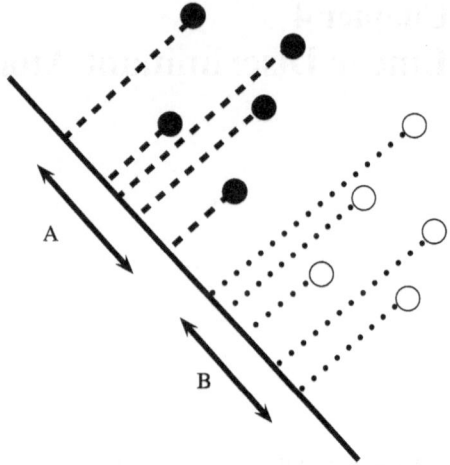

Let $x_1, \ldots, x_p \in \mathbb{R}^m$ be a set of p data samples belonging to two different class sets, A and B. For each class we can define the sample means:

$$\bar{x}_A = \frac{1}{N_A} \sum_{x \in A} x, \; \bar{x}_B = \frac{1}{N_B} \sum_{x \in B} x \tag{4.1}$$

where N_A, N_B are the number of samples in A and B, respectively. Then for each class we can define the positive semidefinite scatter matrices described by the equations:

$$S_A = \sum_{x \in A} (x - \bar{x}_A)(x - \bar{x}_A)^T, \; S_B = \sum_{x \in B} (x - \bar{x}_B)(x - \bar{x}_B)^T \tag{4.2}$$

Each of these matrices expresses the sample variability in each class. Ideally we would like to find a hyperplane, defined by the vector ϕ, for which if we project the data samples their variance would be minimal. That can be expressed as:

$$\min_{\phi} \left(\phi^T S_A \phi + \phi^T S_B \phi \right) = \min_{\phi} \phi^T (S_A + S_B) \phi = \min_{\phi} \phi^T S \phi \tag{4.3}$$

where $S = S_A + S_B$ by definition. On the other side, the scatter matrix between the two classes is given by

$$S_{AB} = (\bar{x}_A - \bar{x}_B)(\bar{x}_A - \bar{x}_B)^T. \tag{4.4}$$

According to Fisher's intuition we wish to find a hyperplane in order to maximize the distance between the means between the two classes and at the same time to minimize the variance in each class. Mathematically this can be described by maximization of Fisher's criterion:

$$\max_{\phi} \mathscr{J}(\phi) = \max_{\phi} \frac{\phi^T S_{AB} \phi}{\phi^T S \phi}. \tag{4.5}$$

This optimization problem can have infinitely many solutions with the same objective function value. That is for a solution ϕ^* all the vectors $c \cdot \phi^*$ give exactly the same value. If, without loss of generality, we replace the denominator with an equality constraint in order to choose only one solution. Then the problem becomes:

$$\max_{\phi} \; \phi^{\mathrm{T}} S_{AB} \phi \tag{4.6a}$$

$$\text{s.t.} \; \phi^{\mathrm{T}} S \phi = 1 \tag{4.6b}$$

The Lagrangian associated with this problem is:

$$\mathscr{L}_{\mathrm{LDA}}(x, \lambda) = \phi^{\mathrm{T}} S_{AB} \phi - \lambda (\phi^{\mathrm{T}} S \phi - 1) \tag{4.7}$$

where λ is the lagrange multiplier that is associated with the constraint (4.6b). Since S_{AB} is positive semidefinite the problem is convex and the global minimum will be at the point for which

$$\frac{\partial \mathscr{L}_{\mathrm{LDA}}(x, \lambda)}{\partial x} = 0 \Leftrightarrow S_{AB} \phi - \lambda S \phi = 0 \tag{4.8}$$

The optimal ϕ can be obtained as the eigenvector that corresponds to the smallest eigenvalue of the following generalized eigensystem:

$$S_{AB} \phi = \lambda S \phi \tag{4.9}$$

Multiclass LDA is a natural extension of the previous case. Given n classes, we need to redefine the scatter matrices: the intra-class matrix becomes

$$S = S_1 + S_2 + \cdots + S_n \tag{4.10}$$

while the inter-class scatter matrix is given by

$$S_{1,\ldots,n} = \sum_{i=1}^{n} p_i (\bar{x}_i - \bar{x})(\bar{x}_i - \bar{x})^{\mathrm{T}} \tag{4.11}$$

where p_i is the number of samples in the i-th class, $\bar{x}_i$ is the mean for each class, and $\bar{x}$ is the total mean vector calculated by

$$\bar{x} = \frac{1}{p} \sum_{i=1}^{n} p_i \bar{x}_i.$$

The linear transformation ϕ we wish to find can be obtained by solving the following generalized eigenvalue problem:

$$S_{1,\ldots,n} \phi = \lambda S \phi.$$

LDA can be used in order to identify which are the most significant features together with the level of significance as expressed by the corresponding coefficient of the projection hyperplane. Also LDA can be used for classifying unknown samples. Once the transformation ϕ is given, the classification can be performed in the transformed space based on some distance measure d. The class of a new point z is determined by

$$\text{class}(z) = \arg\min_{n}\{d(z\phi, \bar{x}_n\phi)\} \tag{4.12}$$

where $\bar{x}_n$ is the centroid of n-th class. This means that first we project the centroids of all classes and the unknown points on the subspace defined by ϕ and we assign the points to the closest class with respect to d.

4.1.1 Generalized Discriminant Analysis

In the case that the linear projection model cannot interpret the data, we need to obtain a nonlinear equivalent of LDA [4]. This can be achieved by the well-studied kernel trick. In this case we embed the original data points (input space) to a higher dimension space (feature space) and then we solve the linear problem. The projection of this linear discriminant in the feature space is a nonlinear discriminant in the input space. This kernel embedding is performed through a function $\kappa: \mathbb{R}^m \mapsto \mathbb{R}^q$ where q is the dimension of the feature space. Then the arithmetic mean on the feature space for each class will be:

$$\bar{x}_1^\kappa = \frac{1}{N_A}\sum_{x\in A}\kappa(x), \ldots, \bar{x}_n^\kappa = \frac{1}{N_B}\sum_{x\in B}\kappa(x), \tag{4.13}$$

the scatter matrices for each class in the feature space

$$V_1 = \sum_{x\in A}(\kappa(x)-\bar{x}_A^\kappa)(\kappa(x)-\bar{x}_A^\kappa)^\mathrm{T}, \ldots, V_n = \sum_{x\in B}(\kappa(x)-\bar{x}_B^\kappa)(\kappa(x)-\bar{x}_B^\kappa)^\mathrm{T} \tag{4.14}$$

and the variance between classes in the feature space will be:

$$B_{1,\ldots,n} = \sum_{i=1}^{n}p_i(\bar{x}_i^\kappa - \bar{x}^\kappa)(\bar{x}_i^\kappa - \bar{x}^\kappa)^\mathrm{T} \tag{4.15}$$

and the Fisher's criterion in the feature space:

$$\min_{y} \mathscr{J}^k(y) = \frac{y^\mathrm{T}(B_{1,\ldots,n})y}{y^\mathrm{T}(\sum_{i=1}^{n}V_i)y} \tag{4.16}$$

The solution can be obtained from the eigenvector that corresponds to the smallest eigenvalue of the generalized eigensystem $B_{1,...,n}y = \lambda \left(\sum_{i=1}^{n} V_i\right) y$. There are several functions that are used as kernel functions in the data mining literature. For a more extensive study of kernel theoretical properties, we refer the reader to [51].

4.2 Robust Discriminant Analysis

The RC formulation of robust LDA was proposed by Kim et al. [30,31]. As in other approaches the motivation for a robust counterpart formulation of LDA comes from the fact that data might be imprecise, thus the means and the standard deviations computed might not be trustworthy estimates of their real values. The approach that we will present here considers the uncertainty on the mean and standard deviation rather on the data points themselves. For the robust case we are interested to determine the optimal value of Fisher's criterion for some undesired, worst case scenario. In terms of optimization this can be described by the following min–max problem.

$$\max_{\phi \neq 0} \min_{\bar{x}_A,\bar{x}_B,S_A,S_B} \frac{\phi^{\mathrm{T}}(\bar{x}_A - \bar{x}_B)(\bar{x}_A - \bar{x}_B)^{\mathrm{T}}\phi}{\phi^{\mathrm{T}}(S_A + S_B)\phi} = \max_{\phi \neq 0} \min_{\bar{x}_A,\bar{x}_B,S_A,S_B} \frac{(\phi^{\mathrm{T}}(\bar{x}_A - \bar{x}_B))^2}{\phi^{\mathrm{T}}(S_A + S_B)\phi} \quad (4.17)$$

In other words we need to estimate the optimal vector ϕ, defining the Fisher's hyperplane, given that a worst case scenario, with respect to means and variances, occurs. This problem's solution strongly depends on the nature of the worst case admissible perturbation set. In general we denote the set of all admissible perturbation $\mathcal{U} \subseteq \mathbb{R}^n \times \mathbb{R}^n \times S_{++}^n \times S_{++}^n$ (by S_{++}^n we denote the set of all positive semidefinite matrices). Then the only constraint of the inner minimization problem would be $(\bar{x}_A,\bar{x}_B,S_A,S_B) \in \mathcal{U}$. In case that we are able to exchange the order of the minimization and the maximization without affecting the problem's structure, we could write:

$$\max_{\phi \neq 0} \min_{(\bar{x}_A,\bar{x}_B,S_A,S_B) \in \mathcal{U}} \frac{(\phi^{\mathrm{T}}(\bar{x}_A - \bar{x}_B))^2}{\phi^{\mathrm{T}}(S_A + S_B)\phi} = \min_{(\bar{x}_A,\bar{x}_B,S_A,S_B) \in \mathcal{U}} (\bar{x}_A - \bar{x}_B)(S_A + S_B)^{-1}(\bar{x}_A - \bar{x}_B)^{\mathrm{T}}$$

$$(4.18)$$

For a general min–max problem, we can write

$$\min_{x \in \mathcal{X}} \max_{y \in \mathcal{Y}} f(x,y) = \max_{y \in \mathcal{Y}} \min_{x \in \mathcal{X}} f(x,y) \quad (4.19)$$

if $f(x,y)$ is convex function with respect to both x, concave with respect to y and also $\mathcal{X}, \mathcal{Y}$ are convex sets. This result is known as strong min–max property and was originally proved by Sion [53]. When convexity does not hold we have the so-called weak min–max property:

$$\min_{x \in \mathcal{X}} \max_{y \in \mathcal{Y}} f(x,y) \geq \max_{y \in \mathcal{Y}} \min_{x \in \mathcal{X}} f(x,y) \quad (4.20)$$

Thus in [30,31] Kim et al. provide such a minimax theorem for the problem under consideration that does not require the strict assumptions of Sion's result. This result is stated in the following theorem

Theorem 4.1. *For the following minimization problem*

$$\min \frac{(w^{\mathrm{T}}a)^2}{w^{\mathrm{T}}Bw} \tag{4.21}$$

let $(a^{\mathrm{opt}}, B^{\mathrm{opt}})$ be the optimal solution. Also let $w^{\mathrm{opt}} = (B^{\mathrm{opt}})^{-1} \cdot a^{\mathrm{opt}}$. Then the point $(w^{\mathrm{opt}}, a^{\mathrm{opt}}, B^{\mathrm{opt}})$ satisfies the following minimax property:

$$\frac{((w^{\mathrm{opt}})^{\mathrm{T}}a^{\mathrm{opt}})^2}{(w^{\mathrm{opt}})^{\mathrm{T}}B^{\mathrm{opt}}w^{\mathrm{opt}}} = \min_{(a,B)}\max_{w} \frac{(w^{\mathrm{T}}a)^2}{w^{\mathrm{T}}Bw} = \max_{w}\min_{(a,B)} \frac{(w^{\mathrm{T}}a)^2}{w^{\mathrm{T}}Bw} \tag{4.22}$$

Proof. See [31].

Here it is worth noting that this result has a variety of applications including signal processing and portfolio optimization (for details, see [30]). Thus the solution for the robust problem can be obtained by solving the following problem:

$$\min \ (\bar{x}_A - \bar{x}_B)(S_A + S_B)^{-1}(\bar{x}_A - \bar{x}_B)^{\mathrm{T}} \tag{4.23a}$$

$$\text{s.t } (\bar{x}_A\bar{x}_B, S_A, S_B) \in \mathscr{U} \tag{4.23b}$$

Assuming that $\mathscr{U}$ is convex problem (4.23) is a convex problem. This holds because the objective function is convex as a matrix fractional function (for detailed proof, see [13]). Next we will examine the robust linear discriminant solution for a special case of uncertainty sets. More specifically let us assume that the Frobenius norm of the differences between the real and the estimated value of the covariance matrices is bounded by a constant. That is

$$\mathscr{U}_S = \mathscr{U}_A \times \mathscr{U}_B \tag{4.24}$$

$$\mathscr{U}_A = \{S_A | \|S_A - \bar{S}_A\|_F \le \delta_A\} \tag{4.25}$$

$$\mathscr{U}_B = \{S_B | \|S_B - \bar{S}_B\| \le \delta_B\} \tag{4.26}$$

In general the worst case minimization problem can be expressed:

$$\min_{(\bar{x}_A, \bar{x}_B, S_A, S_B)} = \min_{(\bar{x}_A, \bar{x}_B) \in \mathscr{U}_x} \frac{(\bar{x}_A - \bar{x}_B)}{\max_{(S_A, S_B) \in \mathscr{U}_S} \phi^{\mathrm{T}}(S_A + S_B)\phi} \tag{4.27}$$

The problem in the denominator can be further simplified:

$$\max_{(S_A, S_B) \in \mathscr{U}_S} \phi^{\mathrm{T}}(S_A + S_B)\phi = \phi^{\mathrm{T}}(\bar{S}_A + \bar{S}_B + \delta_A I + \delta_B I)\phi \tag{4.28}$$

Thus the robust solution will be given by the solution to the convex optimization problem:

$$\min_{(\bar{x}_A, \bar{x}_B)} (\bar{x}_A - \bar{x}_B)^{\mathrm{T}} (\bar{S}_A + \bar{S}_B + \delta_A I + \delta_B I)^{-1} (\bar{x}_A - \bar{x}_B) \qquad (4.29)$$

which is simpler than (4.23). Following similar analysis it is possible to generalize the robust LDA to nonlinear datasets through kernel. Details can be found in [13]. It is worth noting that this connection between regularization and robustness appears over and over in many of the algorithms presented in this monograph. This makes us think that data mining algorithms share some very basic common principles. This can be intuitive but it is very interesting that it is confirmed through mathematically rigorous methods.

Chapter 5
Support Vector Machines

Abstract In this chapter we describe one of the most successful supervised learning algorithms namely suppor vector machines (SVMs). The SVM is one of the conceptually simplest algorithms whereas at the same time one of the best especially for binary classification. Here we illustrate the mathematical formulation of SVM together with its robust equivalent for the most common uncertainty sets.

5.1 Original Problem

Support vector machines (SVM) is one of the most well-known supervised classification algorithms. It was originally proposed by V. Vapnik [60]. The intuition behind the algorithm is that we wish to obtain a hyperplane that "optimally" separates two classes of training data. The power of SVM lies in the fact that it has minimal generalization error (at least in the case of two classes) and the solution can be obtained computationally efficient since it can be formulated as a convex programming problem. Its dual formulation can be used in order to boost the performance even more. As for other supervised classification methods SVM original formulation refers to binary classification problems.

Given a set of data points $x_i, i = 1, \ldots, n$ and an indicator vector $d \in \{-1, 1\}^n$ the class information of the data points we aim to find a hyperplane defined by (w, b) such that the distance between the hyperplane and the closest of the data points of each class (support vectors) (Figs. 5.1 and 5.2). This can be expressed as the following optimization problem:

$$\min_{w,b} \ \frac{1}{2}\|w\|^2 \tag{5.1a}$$

$$\text{s.t.} \ \ d_i\left(w^{\mathrm{T}}x_i + b\right) \geq 1, \ \ i = 1, \ldots, n \tag{5.1b}$$

P. Xanthopoulos et al., *Robust Data Mining*, SpringerBriefs in Optimization,
DOI 10.1007/978-1-4419-9878-1_5,
© Petros Xanthopoulos, Panos M. Pardalos, Theodore B. Trafalis 2013

Fig. 5.1 Separation by
hyperpalnes. The SVM
determines the hyperplane
with the maximum possible
margin

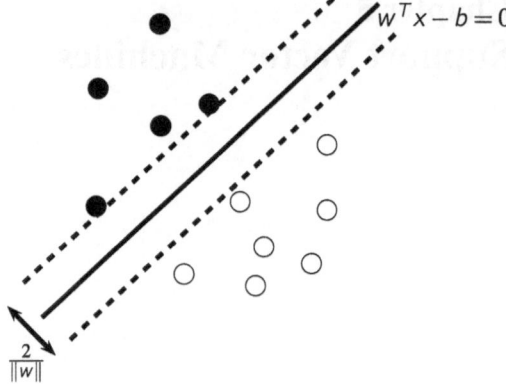

Fig. 5.2 The nonseparable
case. The soft SVM allows
misclassification but
penalizes each misclassified
point

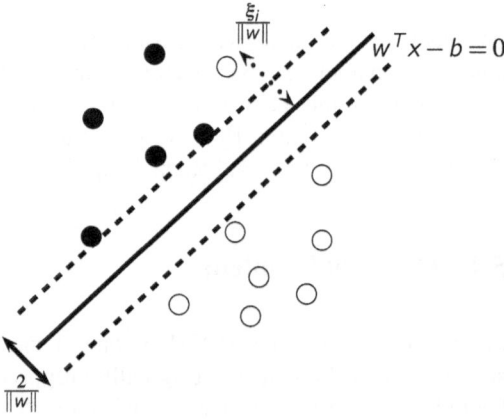

For this problem the Lagrangian equation will be:

$$\mathscr{L}_{\text{SVM}}(w, b, \alpha) = \frac{1}{2} w^{\text{T}} w - \sum_{i=1}^{n} \alpha_i \left[d_i \left(w^{\text{T}} x_i + b \right) - 1 \right] \tag{5.2}$$

where $\alpha = [\alpha_1 \, \alpha_2 \ldots \alpha_n]$ are Lagrange multipliers. In order to determine them we
need to take the partial derivatives with respect to each decision variable and set
them equal to zero.

$$\frac{\partial \mathscr{L}_{\text{SVM}}(w, b, \alpha)}{\partial w} = 0 \Leftrightarrow w = \sum_{i=1}^{n} \alpha_i d_i x_i \tag{5.3a}$$

$$\frac{\partial \mathscr{L}_{\text{SVM}}(w, b, \alpha)}{\partial b} = 0 \Leftrightarrow \sum_{i=1}^{n} \alpha_i d_i = 0 \tag{5.3b}$$

if we substitute in (5.2) we get:

$$\mathscr{L}_{\text{SVM}}(w,b,\alpha) = \frac{1}{2}\sum_{i,j=1}^{n}\alpha_i\alpha_j d_i d_j \langle x_i x_j \rangle - \sum_{i,j=1}^{n}\alpha_i\alpha_j d_j \langle x_j, x_i \rangle + b\sum_{i=1}^{n}\alpha d_i + \sum_{i=1}^{n}\alpha_i$$

(5.4a)

$$= \sum_{i=1}^{n}\alpha_i - \frac{1}{2}\sum_{i,j=1}^{n}\alpha_i\alpha_j d_i d_j \langle x_i, x_j \rangle$$

(5.4b)

Then we can express the dual of the original SMV problem as follows:

$$\max \quad \sum_{i=1}^{n}\alpha_i - \frac{1}{2}\sum_{i,j=1}^{n}\alpha_i\alpha_j d_i d_j \langle x_i, x_j \rangle$$

(5.5a)

$$\text{s.t.} \quad \sum_{i=1}^{n}d_i\alpha_i = 0$$

(5.5b)

$$\alpha_i \geq 0 \qquad\qquad\qquad i = 1,\ldots,n \qquad (5.5c)$$

The last is a also a convex quadratic problem that can be solved efficiently. Once the optimal dual variables $\alpha_i^*, i = 1,\ldots,n$ are found, then the optimal separation hyperplane w^* can be obtained from:

$$w^* = \sum_{i=1}^{n}d_i\alpha_i^* x_i$$

(5.6)

Note that b does not appear in the dual formulation thus it should be estimated through the primal constraints

$$b^* = -\frac{\max_{d_i=-1}\langle w^* x_i \rangle + \min_{d_i=1}\langle w^* x_i \rangle}{2}$$

(5.7)

This model can give a separation hyperplane in case that the two classes are linearly separable. When this assumption does not hold the optimization problem becomes infeasible and we need to slightly modify this original hard margin classification model so that it remains feasible even when some points are misclassified. The idea is to allow misclassified points but at the same time to penalize misclassifications making it a less favorable solution.

$$\min_{w,b,\xi_i} \quad \frac{1}{2}\left(\|w\|^2 + C\sum_{i=1}^{n}\xi_i^2\right)$$

(5.8a)

$$\text{s.t.} \quad d_i\left(w^{\mathsf{T}}x_i + b\right) \geq 1 - \xi_i, \quad i = 1,\ldots,n$$

(5.8b)

where C is the penalization parameter. Note that this model becomes the same as (5.1a) and (5.1b) as $C \mapsto +\infty$. Formulation (5.8a) and (5.8b) is known as soft margin SVM.

This formulation can be seen as a regularized version of formulation (5.1a) and (5.1b). The Lagrangian of (5.8a) and (5.8b) will be:

$$\mathscr{L}_{\text{SVM-S}}(w,b,\xi,\alpha) = \frac{1}{2}w^{\text{T}}w + \frac{C}{2}\sum_{i=1}^{n}\xi_i^2 - \sum_{i=1}^{n}\alpha_i\left[d_i\left(w^{\text{T}}x_i + b - 1 + \xi_i\right)\right] \quad (5.9)$$

where, again, α_i are appropriate Lagrangian multipliers. The dual formulation can be easily obtained in a way similar to the hard margin classifier. The only difference is that now we will have an additional equation associated with the new ξ variables. Setting the derivation of the Lagrangian equal to zero for each of the decision variables gives the following KKT system:

$$\frac{\partial\mathscr{L}_{\text{SVM}}(w,b,\xi,\alpha)}{\partial w} = 0 \Leftrightarrow w = \sum_{i=1}^{n}d_i\alpha_i x_i \quad (5.10a)$$

$$\frac{\partial\mathscr{L}(w,b,\xi,\alpha)}{\partial\xi} = 0 \Leftrightarrow C\xi = \alpha \Leftrightarrow \xi = \frac{1}{C}\alpha \quad (5.10b)$$

$$\frac{\partial\mathscr{L}_{\text{SVM}}(w,b,\xi,\alpha)}{\partial b} = 0 \Leftrightarrow \sum_{i=1}^{n}d_i\alpha_i = 0 \quad (5.10c)$$

Substituting these equation to the primal Lagrangian we obtain:

$$\mathscr{L}_{\text{SVM-S}}(w,b,\xi,\alpha) = \frac{1}{2}w^{\text{T}}w + \frac{C}{2}\sum_{i=1}^{n}\xi_i^2 - \sum_{i=1}^{n}\alpha_i\left[d_i\left(w^{\text{T}}x_i + b - 1 + \xi\right)\right] \quad (5.11a)$$

$$= \sum_{i=1}^{n}\alpha_i - \frac{1}{2}\sum_{i,j=1}^{n}\alpha_i\alpha_j d_i d_j\langle x_i, x_j\rangle + \frac{1}{2C}\langle\alpha,\alpha\rangle - \frac{1}{C}\langle\alpha,\alpha\rangle \quad (5.11b)$$

$$= \sum_{i=1}^{n}\alpha_i - \frac{1}{2}\sum_{i,j=1}^{n}\alpha_i\alpha_j d_i d_j\langle x_i, x_j\rangle - \frac{1}{2C}\langle\alpha,\alpha\rangle \quad (5.11c)$$

$$= \sum_{i=1}^{n}\alpha_i - \frac{1}{2}\sum_{i,j=1}^{n}\alpha_i\alpha_j d_i d_j\left(\langle x_i, x_j\rangle + \frac{1}{C}\delta_{ij}\right) \quad (5.11d)$$

where δ_{ij} is the kronecker δ where it is equal to 1 when $i = j$ and it is zero otherwise. The dual formulation of the problem is thus:

$$\max \sum_{i=1}^{n}\alpha_i - \frac{1}{2}\sum_{i,j=1}^{n}\alpha_i\alpha_j d_i d_j\left(\langle x_i, x_j\rangle + \frac{1}{C}\delta_{ij}\right) \quad (5.12a)$$

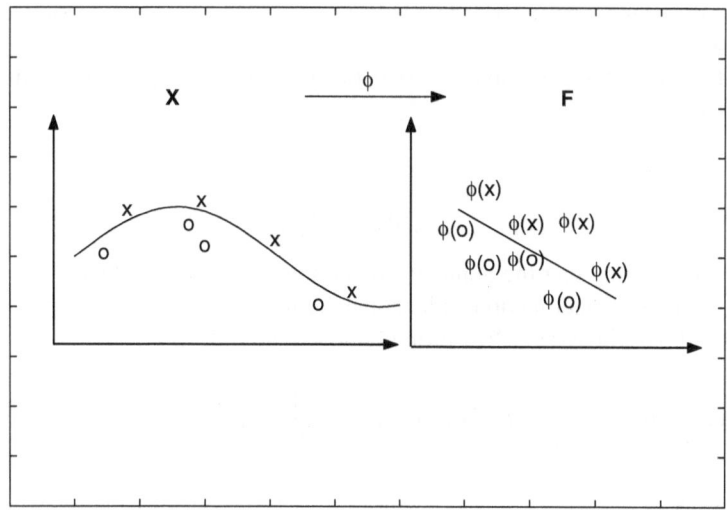

Fig. 5.3 A kernel map converts a nonlinear problem into a linear problem

$$\text{s.t. } \sum_{i=1}^{n} d_i \alpha_i = 0 \tag{5.12b}$$

$$\alpha_i \geq 0 \qquad\qquad i = 1,\ldots,n \tag{5.12c}$$

Once the optimal dual variables have been obtained, the optimal separation hyperplane can be recovered similar as in hard margin classifier. Given the hyperplane a new point x_u can be classified in one of the two classes based on the following rule:

$$d_{x_u} = \text{sgn}(w^{\text{T}} x_u + b) \tag{5.13}$$

where $\text{sgn}(\cdot)$ is the sign function. To address the problem of nonlinearity that frequently occurs in real world problems, one can use kernel methods. Kernel methods [50] provide an alternative approach by mapping data points x in the input space into a higher dimensional feature space F through a map φ such that $\varphi : x \mapsto \varphi(x)$. Therefore a point x in the input space becomes $\varphi(x)$ in the feature space.

Even though very often the function $\varphi(x)$ is not available, cannot be computed, or does not even exist, the dot product $\langle \varphi(x_1), \varphi(x_2) \rangle$ can still be computed in the feature space through a kernel function. In order to employ the kernel method, it is necessary to express the separation constraints in the feature space in terms of inner products between the data points $\varphi(x_i)$. Then in the higher dimensional feature space we can construct a linear decision function that represents a nonlinear decision function in the input space. Figure 5.3 describes an example of a kernel mapping from a two-dimensional input space to a two-dimensional feature space. In the input space the data cannot be separated linearly, however can be linearly separated in the feature space.

The following three nonlinear kernel functions are usually used in the SVM literature [24]:

- polynomial: $(x^{\mathrm{T}}x + 1)^p$,
- radial basis function (RBF): $\exp(-\frac{1}{2\sigma^2}\|x - x_i\|^2)$,
- tangent hyperbolic (sigmoid): $\tanh(\beta x^{\mathrm{T}}x + \beta_1)$, where $\beta, \beta_1 \in \mathbb{R}$.

It is worth noting that the nonlinear version of SVM can be obtained if we just replace the dot product function with another kernel function $\varphi(x_i, x_j)$. For example the dual soft margin kernel SVM formulation will be:

$$\max \; \sum_{i=1}^{n} \alpha_i - \frac{1}{2} \sum_{i,j=1}^{n} \alpha_i \alpha_j d_i d_j \left(\varphi(x_i, x_j) + \frac{1}{C}\delta_{ij} \right) \tag{5.14a}$$

$$\text{s.t.} \; \sum_{i=1}^{n} d_i \alpha_i = 0 \tag{5.14b}$$

$$\alpha_i \geq 0 \qquad\qquad\qquad\qquad i = 1, \dots, n \tag{5.14c}$$

In fact the linear case is a special kernel case as the dot product can be seen as an admissible kernel function, i.e. $\varphi(\cdot, \cdot) = \langle \cdot, \cdot \rangle$. One of the fundamental limitations of the generic formulation of soft SVM is that it is proposed just for the two-class case (binary classification). This might pose a problem, since many of the real world problems involve data that belong to more than two classes.

Majority voting scheme [60]:   According to this approach, given a total of N classes, we solve the SVM problem for all binary combinations (pairs) of classes. For example for a three-class problem (class A, class B, and class C), we find the separation hyperplanes that correspond to the problems A vs B, A vs C, and B vs C. When a new point comes, then each classifier "decides" on the class of this point. Finally the point is classified into the class with the most "votes."

Directed acyclic graph approach [42]:   For the majority voting process one needs to construct a large number of training binary classifiers in order to infer the class of an unknown sample. This can pose a computational problem to the performance. Thus in the directed acyclic graph we try to minimize the number of necessary classifiers required. This can be achieved by considering a tree that eliminates one class at each level. An example with four classes is illustrated in Fig. 5.4.

A straightforward observation regarding these two multiclass generalization strategies is that they can be used for any type of binary classifiers (not only SVM) with or without the use of kernel.

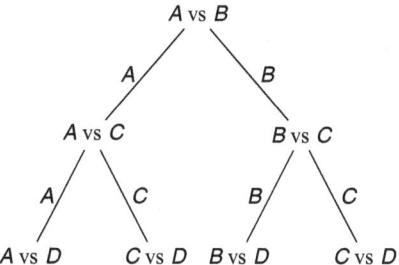

Fig. 5.4 Directed acyclic graph approach for a four-class example (A,B,C and D). At the first level of the tree the sample goes through the A vs B classifier. Then depending on the outcome the sample is tested through the A vs C or B vs C. The total number of binary classifiers needed equals the depth of the tree

5.1.1 Alternative Objective Function

In a more general framework there has been proposed alternative objective function that can be used for SVM classification yielding in computationally different problems. In the general form one can express the objective function as $f_p(w)$ where p corresponds to the type of norm. Along these lines we can express the penalty function as $g(C,\xi)$ where C can be either a number, meaning that all points are penalized the same way, or a diagonal matrix, with each diagonal element corresponding to the penalization coefficient of each data sample. In practice, different penalization coefficient might be useful when, for example, the classification problem is imbalanced (the training samples of one class is much higher than the other). For the case of SVM that we already presented we assumed quadratic objective and penalty functions:

$$f_2(w) = \|w\|_2^2 = w^T w, \quad g(C,\xi) = \xi^T C \xi \tag{5.15}$$

Another popular choice is

$$f(w) = \|w\|_1, \quad g(C,\xi) = \langle C,\xi \rangle \tag{5.16}$$

In this case the SVM formulation becomes:

$$\min \; \|w\|_1 + \langle C,\xi \rangle \tag{5.17a}$$

$$\text{s.t.} \; d_i(w^T x_i + b) \geq 1 - \xi_i, \quad i = 1,\dots,n \tag{5.17b}$$

It is easy to show that the last formulation can be solved as a linear program (LP). More specifically if we introduce the axillary variable α we can obtain the following equivalent formulation of problem (5.17a) and (5.17b)

$$\min \ \sum_{i=1}^{n} \alpha_i + \langle C, \xi \rangle \tag{5.18a}$$

$$\text{s.t. } d_i \left(\sum_{i=1}^{n} \alpha_i \langle x_i, x_j \rangle + b \right) \geq 1 - \xi_i, \ i = 1, \ldots, n \tag{5.18b}$$

$$\alpha_i \geq 0, \ \xi_i \geq 0, \ i = 1, \ldots, n \tag{5.18c}$$

It is worth noting that the linear programming approach was developed independently from the quadratic one.

5.2 Robust Support Vector Machines

The SVMs is one of the most well-studied application of robust optimization in data mining. The theoretical and practical issues have been extensivaly explored through the works of Trafalis et al. [56, 58], Nemirovski et al. [6], and Xu et al. [62]. It is of particular interest that robust SVM formulations are tractable for a variety of perturbation sets. At the same time there is clear theoretical connection between particular robustification and regularization [62]. On the other side, several robust optimization formulation can be solved as conic problems. If we recall the primal soft margin SVM formulation presented in the previous section:

$$\min_{w,b,\xi_i} \ \frac{1}{2} \left(\|w\|^2 + C \sum_{i=1}^{n} \xi_i^2 \right) \tag{5.19a}$$

$$\text{s.t. } d_i \left(w^\mathsf{T} x_i + b \right) \geq 1 - \xi_i, \ i = 1, \ldots, n \tag{5.19b}$$

$$\xi_i \geq 0, \ i = 1, \ldots, n \tag{5.19c}$$

for the robust case we replace each point x_i with $\tilde{x}_i = \bar{x}_i + \sigma_i$ where $\bar{x}_i$ are the nominal (known) values and σ_i is an additive unknown perturbation that belongs to a well-defined uncertainty set. The objective is to solve the problem for the worst case perturbation. Thus the general robust optimization problem formulation can be stated as follows:

$$\min_{w,b,\xi_i} \ \frac{1}{2} \left(\|w\|^2 + C \sum_{i=1}^{n} \xi_i^2 \right) \tag{5.20a}$$

$$\text{s.t. } \min_{\sigma_i} \left(d_i \left(w^\mathsf{T} (\bar{x}_i + \sigma_i) + b \right) \right) \geq 1 - \xi_i, \ i = 1, \ldots, n \tag{5.20b}$$

$$\xi \geq 0, i = 1, \ldots, n \tag{5.20c}$$

Note that since the expression of constraint (5.20b) corresponds to the distance of the ith point to the separation hyperplane the worst case σ_i would be the one that minimizes this distance. An equivalent form of constraint (5.20b) is:

$$d_i \left(w^{\mathrm{T}} \bar{x}_i + b \right) + \min_{\sigma_i} d_i \left(w^{\mathrm{T}} \sigma_i \right) \geq 1 - \xi_i, \quad i = 1, \ldots, n \tag{5.21}$$

Thus, solving the robust SVM optimization problem involves the following problem:

$$\min_{\sigma_i \in \mathscr{U}_{\sigma_i}} d_i \left(w^{\mathrm{T}} \sigma_i \right), \quad i = 1, \ldots, n \tag{5.22}$$

for fixed w, where $\mathscr{U}_{\sigma_i}$ is the sets of admissible perturbations corresponding to ith sample. Suppose that the l_p norm of the unknown perturbations are bounded by known constant.

$$\min \ d_i \left(w^{\mathrm{T}} \sigma_i \right), \quad i = 1, \ldots, n \tag{5.23a}$$

$$\text{s.t.} \ \|\sigma_i\|_p \leq \rho_i \tag{5.23b}$$

By using Hölders inequality (see appendix) we can obtain:

$$|d_i(w^{\mathrm{T}} \sigma_i)| \leq \|w\|_q \|\sigma_i\|_p \leq \rho_i \|w\|_q \tag{5.24}$$

where $\| \cdot \|_q$ is the dual norm of $\| \cdot \|_p$. Equivalently we can obtain:

$$-\rho_i \|w\|_q \leq d_i(w^{\mathrm{T}} \sigma_i) \tag{5.25}$$

Thus the minimum of this expression will be $-\rho_i \|w\|_q$. If we substitute this expression in the original problem, we obtain:

$$\min_{w,b,\xi_i} \ \frac{1}{2} \left(\|w\|^2 + C \sum_{i=1}^{n} \xi_i^2 \right) \tag{5.26a}$$

$$\text{s.t.} \ d_i \left(w^{\mathrm{T}} (\bar{x}_i + \sigma_i) + b \right) - \rho_i \|w\|_q \geq 1 - \xi_i, \quad i = 1, \ldots, n \tag{5.26b}$$

$$\xi_i \geq 0, \quad i = 1, \ldots, n \tag{5.26c}$$

The structure of the obtained optimization problem depends on the norm p. Next we will present some "interesting" case. It is easy to determine the value of q from $1/p + 1/q = 1$ (for details see appendix). For $p = q = 2$, we obtain the following formulation:

$$\min_{w,b,\xi_i} \ \frac{1}{2} \left(\|w\|^2 + C \sum_{i=1}^{n} \xi_i^2 \right) \tag{5.27a}$$

$$\text{s.t.} \ d_i \left(w^{\mathrm{T}} \bar{x}_i + b \right) - \rho_i \|w\|_2 \geq 1 - \xi_i, \quad i = 1, \ldots, n \tag{5.27b}$$

$$\xi_i \geq 0, \quad i = 1, \ldots, n \tag{5.27c}$$

The last formulation can be seen as a regularization of the original problem. Another interesting case is when the uncertainty is described with respect to the first norm (box constraints). In this case the robust formulation will be:

$$\min_{w,b,\xi_i} \frac{1}{2}\left(\|w\|_\infty + C\sum_{i=1}^{n}\xi_i\right) \tag{5.28a}$$

$$\text{s.t. } d_i\left(w^{\mathrm{T}}(\bar{x}_i + \sigma_i) + b\right) - \rho_i\|w\|_\infty \geq 1 - \xi_i, \quad i = 1,\ldots,n \tag{5.28b}$$

$$\xi_i \geq 0, \quad i = 1,\ldots,n \tag{5.28c}$$

Since the dual of l_1 norm is the l_∞ norm. If we further more assume that the norm of the loss function is expressed with respect to the l_1 norm, then the obtained optimization problem can be solved as a linear program (LP). The drawback of this formulation is that it is not kernalizable. More specifically if we introduce the axillary variable α, we can obtain the following equivalent formulation of problem (5.28a), (5.28b) and (5.28c):

$$\min_{\alpha,w,b,\xi_i} \alpha + \langle C,\xi\rangle \tag{5.29a}$$

$$\text{s.t. } d_i\left(w^{\mathrm{T}}(\bar{x}_i + \sigma_i) + b\right) - \rho_i\alpha \geq 1 - \xi_i \qquad i = 1,\ldots,n \tag{5.29b}$$

$$\xi_i \geq 0 \qquad i = 1,\ldots,n \tag{5.29c}$$

$$\alpha \geq -w_k \qquad k = 1,\ldots,n \tag{5.29d}$$

$$\alpha \geq w_k \qquad k = 1,\ldots,n \tag{5.29e}$$

$$\alpha \geq 0 \tag{5.29f}$$

If the perturbations are expressed with respect to the l_∞ norm, then the equivalent formulation of SVM is:

$$\min \; (\|w\|_1 + \langle C,\xi\rangle) \tag{5.30a}$$

$$\text{s.t. } d_i\left(w^{\mathrm{T}}x_i + b\right) - \rho_i\|w\|_1 \geq 1 - \xi_i \qquad i = 1,\ldots,n \tag{5.30b}$$

$$\xi_i \geq 0 \qquad i = 1,\ldots,n \tag{5.30c}$$

In the same way if we introduce the auxiliary variables $\alpha_1, \alpha_2, \ldots, \alpha_n$, the formulation becomes

$$\min_{\alpha_i,w,b,\xi_i} \sum_{i=1}^{n}\alpha_i + \langle C,\xi\rangle \tag{5.31a}$$

$$\text{s.t. } d_i\left(w^{\mathrm{T}}\bar{x} + b\right) - \rho_i\sum_{i=1}^{n}\alpha_i \geq 1 - \xi_i \qquad i = 1,\ldots,n \tag{5.31b}$$

$$\xi_i \geq 0 \qquad\qquad i=1,\ldots,n \qquad (5.31\text{c})$$

$$\alpha_i \geq -w_i \qquad\qquad i=1,\ldots,n \qquad (5.31\text{d})$$

$$\alpha_i \geq w_i \qquad\qquad i=1,\ldots,n \qquad (5.31\text{e})$$

$$\alpha_i \geq 0 \qquad\qquad i=1,\ldots,n \qquad (5.31\text{f})$$

It is worth noting that for all robust formulations of SVM the classification rule remains the same as for the nominal case: $\text{class}(u) = \text{sgn}\left(w^{\mathrm{T}}u+b\right)$. Next we will describe the feasibility approach formulation, an SVM like optimization approach with linear objective function, and its robust equivalent.

5.3   Feasibility-Approach as an Optimization Problem

As the SVM algorithm, the feasibility-approach algorithm can be formulated through an optimization problem. Suppose that we have a set of ℓ samples $\{x_1, x_2, \ldots, x_\ell\}$ and we want a weight vector w and a bias b that satisfies $y_i(w^{\mathrm{T}}x_i + b) \geq 1$ for all $i = 1, \ldots, \ell$. This feasibility problem can be expressed as an LP problem [19] by introducing an artificial variable $t \geq 0$ and solving the following

$$\min\ t$$

$$\text{s.t.}\ \ y_i(w^{\mathrm{T}}x_i+b)+t \geq 1$$

$$t \geq 0, \qquad\qquad (5.32)$$

where $w \in \mathbb{R}^n$ and b and t are scalar variables. By minimizing the slack variable t we can decide if the separation is feasible. If the optimal value $\hat{t} = 0$, then the samples are linearly separable and we have a solution. If $\hat{t} > 0$, there is no separating hyperplane and we have a proof that the samples are nonseparable. In contrast to the SVM approach, we keep the same slack variable t constant for each separation constraint.

5.3.1   Robust Feasibility-Approach and Robust SVM Formulations

In [48] Santosa and Trafalis proposed the robust counterpart algorithm of the feasibility approach formulation. Consider that our data are perturbed. Instead of having the input data point x_i, now we have $x_i = \tilde{x}_i + u_i$ where u_i is a bounded perturbation with $\|u_i\| \leq \sqrt{\eta}$, η is a positive number, and $\tilde{x}_i$ is the center of the

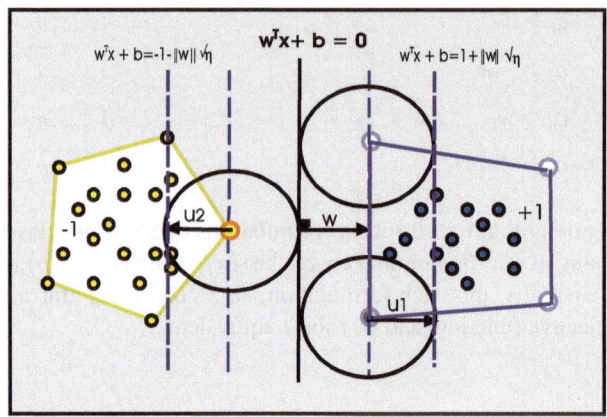

Fig. 5.5 Finding the best classifier for data with uncertainty. The bounding planes are moved to the edge of the spheres to obtain maximum margin

uncertainty sphere where our data point is located. Therefore, the constraints in (5.32) become

$$y_i(\langle w, x_i \rangle + b) + t \geq 1$$

$$\Leftrightarrow y_i(\langle w, \tilde{x}_i \rangle + \langle w, u_i \rangle + b) + t \geq 1, i = 1, \ldots, \ell$$

$$t \geq 0,$$

$$\|u_i\| \leq \sqrt{n}. \tag{5.33}$$

Our concern is the problem of classification with respect to two classes and for every realization of u_i in the sphere $S(0, \sqrt{\eta})$. In order to increase the margin between the two classes (and therefore having the best separating hyperplane), we try to minimize the dot product of w and u_i in one side of the separating hyperplane (class -1) and maximize the dot product of w and u_i in the other side (class 1) subject to $\|u_i\| \leq \sqrt{\eta}$. In other words in (5.33) we replace $\langle w, u_i \rangle$ with its minimum value for the negative examples (class -1) and with its maximum value for the positive examples (class 1). By this logic, we are trying to maximize the distance between the classifier and points on different classes (see Fig. 5.5) and therefore increasing the margin of separation.

Therefore we have to solve the following two problems

$$\max \ \langle w, u_i \rangle \ \text{s.t.} \ \|u_i\| \leq \sqrt{\eta}$$

for $y_i = +1$ and

$$\min \ \langle w, u_i \rangle \ \text{s.t.} \ \|u_i\| \leq \sqrt{\eta}$$

for $y_i = -1$.

Using Cauchy–Schwarz inequality, the maximum and the minimum of the dot product of $\langle w, u_i \rangle$ will be $\sqrt{\eta}\,\|w\|$ and $-\sqrt{\eta}\,\|w\|$ respectively. By substituting the maximum value of $\langle w, u_i \rangle$ for $y_i = 1$ and its minimum value for $y_i = -1$ in (5.33), we have

$$\text{min } t$$
$$\text{s.t. } \sqrt{\eta}\,\|w\| + w^{\mathrm{T}}\tilde{x}_i + b + t \geq 1, \text{for } y_i = +1$$
$$\sqrt{\eta}\,\|w\| - w^{\mathrm{T}}\tilde{x}_i - b + t \geq 1, \text{for } y_i = -1$$
$$t \geq 0. \tag{5.34}$$

If we map the data from the input space to the feature space F, then $g(x) = \text{sign}(w^{\mathrm{T}}\varphi(x) + b)$ is a decision function in the feature space. In the feature space, (5.34) becomes

$$\text{min } t$$
$$\text{s.t. } \sqrt{\eta}\,\|w\| + w^{\mathrm{T}}\varphi(\tilde{x}_i) + b + t \geq 1, \text{for } y_i = +1$$
$$\sqrt{\eta}\,\|w\| - w^{\mathrm{T}}\varphi(\tilde{x}_i) - b + t \geq 1, \text{for } y_i = -1$$
$$t \geq 0. \tag{5.35}$$

We can represent w as

$$w = \sum_{i=1}^{\ell} \alpha_i \varphi(\tilde{x}_i), \tag{5.36}$$

where $\alpha_i \in \mathbb{R}$. By substituting w with the above representation and substituting $\langle \varphi(\tilde{x}), \varphi(\tilde{x}) \rangle$ with K, we have the following robust feasibility-approach formulation

$$\text{min } t$$
$$\text{s.t. } \sqrt{\eta}\sqrt{\alpha^{\mathrm{T}}K\alpha} + K_i\alpha + b + t \geq 1, \text{for } y_i = +1$$
$$\sqrt{\eta}\sqrt{\alpha^{\mathrm{T}}K\alpha} - K_i\alpha - b + t \geq 1, \text{for } y_i = -1$$
$$t \geq 0, \tag{5.37}$$

where K_i is the $1 \times \ell$ vector corresponding to the ith line of the kernel matrix K. Note that we reorder the rows of the matrix K based on the label. It is important to note that most of the time we do not need to know explicitly the map φ. The important idea is that we can replace $\langle \varphi(x), \varphi(x) \rangle$ with any suitable kernel $k(x,x)$.

By modifying the constraints of the SVM model incorporating noise as in the feasibility-approach, we have the following robust SVM model formulation:

$$\min \ \frac{1}{2}\alpha^{\mathrm{T}}K\alpha + C\sum_{i=1}^{\ell} t_i$$

$$\text{s.t.} \ \sqrt{\eta}\sqrt{\alpha^{\mathrm{T}}K\alpha} - K_i\alpha - b + t_i \geq 1, \text{for } y_i = -1$$

$$\sqrt{\eta}\sqrt{\alpha^{\mathrm{T}}K\alpha} + K_i\alpha + b + t_i \geq 1, \text{for } y_i = +1$$

$$t_i \geq 0. \tag{5.38a}$$

Note that the above formulations are SOCP problems. By margin(η), we define the margin of separation when the level of uncertainty is η. Then

$$\text{margin}(\eta) = \frac{(1 + \|w\|\sqrt{\eta} - b) - (-1 - b + \sqrt{(\eta)}\|w\|)}{\|w\|}$$

$$= \frac{2 + 2\sqrt{\eta}\|w\|}{\|w\|} = \frac{2}{\|w\|} + 2\sqrt{\eta} = \text{margin}(0) + 2\sqrt{\eta} \tag{5.39}$$

The above equation shows that as we increase the level of uncertainty η, the margin is increasing in contrast to [57] formulation where the margin is decreasing.

Chapter 6
Conclusion

In this work, we presented some of the major recent advances of robust optimization in data mining. Through this monograph, we examined most of the data mining methods from the scope of uncertainty handling with only exception the principal component analysis (PCA) transformation. Nevertheless the uncertainty can be seen as a special case of prior knowledge. In prior knowledge classification, for example, we are given together with the training sets some additional information about the input space. Another type of prior knowledge other than uncertainty is the so-called expert knowledge, e.g., binary rule of the type "if feature a is more than M_1 and feature b less than M_2 then the sample belongs to class x." There has been significant amount of research in the area of prior knowledge classification [33, 49] but there has not been a significant study of robust optimization on this direction.

On the other side there have been several other methods able to handle uncertainty like stochastic programming as we already mentioned at the beginning of the manuscript. Some techniques, for example, conditional value at risk (CVAR), have been extensively used in portfolio optimization and in other risk related decision systems optimization problems [46] but their value for machine learning has not been fully investigated.

Application of robust optimization in machine learning would be an alternative method for data reduction. In this case we could replace groups of points by convex shapes, such as balls, squares or ellipsoids, that enclose them. Then the supervised learning algorithm can be trained just by considering these shapes instead of the full sets of points.

P. Xanthopoulos et al., *Robust Data Mining*, SpringerBriefs in Optimization,
DOI 10.1007/978-1-4419-9878-1_6,
© Petros Xanthopoulos, Panos M. Pardalos, Theodore B. Trafalis 2013

Appendix A
Optimality Conditions

Here we will briefly discuss the Karush–Kuhn–Tucker (KKT) Optimality Conditions and the method of Lagrange multipliers that is extensively used through this work. In this section, for the sake of completion we are going to describe the technical details related to optimality of convex programs and the relation with KKT systems and methods of Lagrange multipliers. First we will start by giving some essential definitions related to convexity. First we give the definition of a convex function and convex set.

Definition A.1. A function $f : X \subseteq \mathbb{R}^m \mapsto \mathbb{R}$ is called convex when $\lambda f(x) + (1 - \lambda) f(x) \geq f(\lambda x + (1 - \lambda)x)$ for $0 \leq \lambda \leq 1$ and $\forall x \in X$.

Definition A.2. A set X is called convex when for any two points $x_1, x_2 \in X$ the point $\lambda x_1 + (1 - \lambda)x_2 \in X$ for $0 \leq \lambda \leq 1$.

Now we are ready to define a convex optimization problem

Definition A.3. An optimization problem $\min_{x \in X} f(x)$ is called convex when $f(x)$ is a convex function and X is a convex set.

The class of convex problems is really important because they are classified as problems that are computationally tractable. This allows the implementation of fast algorithms for data analysis methods that are realized as convex problems. Processing of massive datasets can be realized because of this property. Once we have defined the convex optimization problem in terms of the properties of its objective function and its feasible region we will state some basic results related to their optimality.

Corollary A.1. *For a convex minimization problem a local minimum x^* is always a global minimum as well. That is if $f(x^*) \leq (f(x))$ for $x \in S$ where $S \subseteq X$ then $f(x^*) \leq f(x)$ for $x \in X$.*

P. Xanthopoulos et al., *Robust Data Mining*, SpringerBriefs in Optimization,
DOI 10.1007/978-1-4419-9878-1,
© Petros Xanthopoulos, Panos M. Pardalos, Theodore B. Trafalis 2013

Proof. Let x^* be a local minimum such that $f(x^*) < f(x), x \in S \subseteq X$ and another point $\bar{x}$ being the global minimum such that $f(\bar{x}) < f(x), x \in X$. Then by convexity of the objective function it holds that

$$f(\lambda \bar{x} + (1 - \lambda)x^*) = f(x^* + \lambda(\bar{x} - x^*)) \leq \lambda f(\bar{x}) + (1 - \lambda)f(x^*) < f(x^*) \quad (A.1)$$

on the other side by local optimality of point $\bar{x}$ we have that there exist $\lambda^* > 0$ such that

$$f(x^*) \leq f(x^* + \lambda(\bar{x} - x^*)), \, 0 \leq \lambda \leq \lambda^* \quad (A.2)$$

which is a contradiction.

This is an important consequence that explains in part the computational track-tability of convex problems. Next we define the critical points that are extremely important for the characterization of global optima of convex problems. But before that we need to introduce the notion of extreme directions.

Definition A.4. A vector $d\mathbb{R}^n$ is called feasible direction with respect to a set S at a point x if there exist $c \in \mathbb{R}$ such that $x + \lambda d \in S$ for every $0 < \lambda < c$.

Definition A.5. For a convex optimization problem $\min_{x \in X} f(x)$ where f differentiable every point that satisfies $d^T \nabla f(x^*) \geq 0, d \in Z(x^*)$ (where $Z(x^*)$ is the set of all feasible directions of the point x^*) is called a critical (or stationary) point.

Critical points are very important in optimization as they are used in order to characterize local optimality in general optimization problems. In a general differentiable setup stationary points characterize local minima. This is formalized through the following theorem.

Theorem A.1. *If x^* is a local minimum of a continuously diffentiable function f defined on a convex set S, then it satisfies $d^T \nabla f(x^*) \geq 0, d \in Z(x^*)$.*

Proof. [25] p. 14.

Due to the specific properties of convexity, in convex programming, critical points are used in order to characterize global optimal solutions as well. This is stated through the following theorem.

Theorem A.2. *if f is a continuously differentiable function on an open set containing S, and S is a convex set then $x^* \in S$ is a global minimum if and only if x^* is a stationary point.*

Proof. [25] pp. 14–15.

The last theorem is a very strong result that connects stationary points with global optimality. Since stationary points are so important for solving convex optimization problems, it is also important to establish a methodology that would allow us to discover such points. This is exactly the goal of Karush–Kuhn–Tucker conditions and method of Lagrangian multipliers. (They are actually different sides of the same coin.) This systematic methodology was first introduced by Lagrange in 1797 and it was generalized through the master thesis of Karush [29] and finally they became more popular known through the work of Kuhn and Tucker [32]. These conditions are formally stated through the next theorem.

Theorem A.3 (KKT conditions). *Given the following optimization problem*

$$\min \quad f(x) \tag{A.3a}$$

$$\text{s.t.} \quad g_i(x) \geq 0, \quad i = 1,\dots,n \tag{A.3b}$$

$$\phantom{\text{s.t.} \quad} h_i(x) = 0, \quad i = 1,\dots,m \tag{A.3c}$$

$$\phantom{\text{s.t.} \quad} x \geq 0 \tag{A.3d}$$

The following conditions (KTT) are necessary for optimality

$$\nabla f(x^*) + \sum_{i=1}^{n} \lambda_i \nabla g_i(x^*) + \sum_{i=1}^{m} \mu_i \nabla h_i(x^*) \tag{A.4a}$$

$$\lambda_i g_i(x^*) = 0 \qquad\qquad\qquad i = 1,\dots,n \tag{A.4b}$$

$$\lambda_i \geq 0 \qquad\qquad\qquad\quad\; i = 1,\dots,n \tag{A.4c}$$

For the special case that $f(\cdot), g(\cdot), h(\cdot)$ are convex functions, then the KKT conditions are also sufficient for optimality.

Proof. See [25]

The (A.4a) is also known as Lagrangian equation and λ_i are also known as lagrange multipliers. Thus one can determine stationary for a problem by just finding the roots of the Lagrangian's first derivative. For the general case this method is formalized through the Karush–Kuhn–Tucker optimality conditions. The important of these conditions is that under convexity assumptions they are necessary and sufficient.

Due to the aforementioned results that connect stationary point with optimality we can clearly see that one can solve a convex optimization problem just by solving the corresponding KKT system. The corresponding points would be the solution to the original problem.

Appendix B
Dual Norms

Dual norms is a mathematical tool, necessary for the analysis of robust support vector machines formulation.

Definition B.1. For a norm $\| \cdot \|$ we define the dual norm $\| \cdot \|_*$ as follows

$$\|x\|_* = \sup\{x^\mathrm{T}\alpha | \|x\| \leq \alpha\} \tag{B.1}$$

There are several properties associated with the dual norm that we will briefly discuss here.

Property B.1. A dual norm of a dual norm is the original norm itself. In other words

$$\|x\|_{**} = \|x\| \tag{B.2}$$

Property B.2. A dual of an l_a norm is l_b norm where a and b satisfy the following equation

$$\frac{1}{a} + \frac{1}{b} = 1 \Leftrightarrow b = \frac{a}{a-1} \tag{B.3}$$

Immediate results of the previous property is that

- The dual norm of the Euclidean norm is the Euclidean norm $(b = 2/(2-1) = 2)$.
- The dual norm of the l_1 norm is l_∞

Next we will state Hölders inequality and Cauchy Swartz inequality which are two fundamental inequalities that connect the primal and the dual norm.

Theorem B.1 (Hölders inequality). *For a pair of dual norms a and b, the following inequality holds:*

$$\langle x \cdot y \rangle \leq \|x\|_a \|y\|_b \tag{B.4}$$

For the special case that $a = b = 2$ then Hölders inequality reduces to Cauchy–Swartz inequality

$$\langle x \cdot y \rangle \leq \|x\|_2 \|y\|_2 \tag{B.5}$$

P. Xanthopoulos et al., *Robust Data Mining*, SpringerBriefs in Optimization, 55
DOI 10.1007/978-1-4419-9878-1,
© Petros Xanthopoulos, Panos M. Pardalos, Theodore B. Trafalis 2013

References

1. Abello, J., Pardalos, P., Resende, M.: Handbook of massive data sets. Kluwer Academic Publishers Norwell, MA, USA (2002)
2. Angelosante, D., Giannakis, G.: RLS-weighted Lasso for adaptive estimation of sparse signals. In: Acoustics, Speech and Signal Processing, 2009. ICASSP 2009. IEEE International Conference on, pp. 3245–3248. IEEE (2009)
3. d Aspremont, A., El Ghaoui, L., Jordan, M., Lanckriet, G.: A direct formulation for sparse pca using semidefinite programming. SIAM review **49**(3), 434 (2007)
4. Baudat, G., Anouar, F.: Generalized discriminant analysis using a kernel approach. Neural computation **12**(10), 2385–2404 (2000)
5. Bayes, T.: An essay towards solving a problem in the doctrine of chances. R. Soc. Lond. Philos. Trans **53**, 370–418 (1763)
6. Ben-Tal, A., El Ghaoui, L., Nemirovski, A.S.: Robust optimization. Princeton Univ Pr (2009)
7. Ben-Tal, A., Nemirovski, A.: Robust solutions of linear programming problems contaminated with uncertain data. Mathematical Programming **88**(3), 411–424 (2000)
8. Bertsimas, D., Pachamanova, D., Sim, M.: Robust linear optimization under general norms. Operations Research Letters **32**(6), 510–516 (2004)
9. Bertsimas, D., Sim, M.: The price of robustness. Operations Research **52**(1), 35–53 (2004)
10. Birge, J., Louveaux, F.: Introduction to stochastic programming. Springer Verlag (1997)
11. Bishop, C.: Pattern recognition and machine learning. Springer New York (2006)
12. Blondin, J., Saad, A.: Metaheuristic techniques for support vector machine model selection. In: Hybrid Intelligent Systems (HIS), 2010 10th International Conference on, pp. 197–200 (2010)
13. Boyd, S., Vandenberghe, L.: Convex optimization. Cambridge Univ Pr (2004)
14. Bratko, I.: Prolog programming for artificial intelligence. Addison-Wesley Longman Ltd (2001)
15. Bryson, A., Ho, Y.: Applied optimal control: optimization, estimation, and control. Hemisphere Pub (1975)
16. Calderbank, R., Jafarpour, S.: Reed Muller Sensing Matrices and the LASSO. Sequences and Their Applications–SETA 2010 pp. 442–463 (2010)
17. Chandrasekaran, S., Golub, G., Gu, M., Sayed, A.: Parameter estimation in the presence of bounded modeling errors. Signal Processing Letters, IEEE **4**(7), 195–197 (1997)
18. Chandrasekaran, S., Golub, G., Gu, M., Sayed, A.: Parameter estimation in the presence of bounded data uncertainties. SIAM Journal on Matrix Analysis and Applications **19**(1), 235–252 (1998)
19. Chvatal, V.: Linear Programming. Freeman and Company (1983)
20. El Ghaoui, L., Lebret, H.: Robust solutions to least-squares problems with uncertain data. SIAM Journal on Matrix Analysis and Applications **18**, 1035–1064 (1997)

21. Fan, N., Pardalos, P.: Robust optimization of graph partitioning and critical node detection in analyzing networks. Combinatorial Optimization and Applications pp. 170–183 (2010)
22. Fan, N., Zheng, Q., Pardalos, P.: Robust optimization of graph partitioning involving interval uncertainty. Theoretical Computer Science (2011)
23. Fisher, R.: The use of multiple measurements in taxonomic problems. Annals of Eugenics **7**(7), 179–188 (1936)
24. Haykin, S.: Neural Network: A Comprehensive Foundation. Prentice Hall, New Jersey (1999)
25. Horst, R., Pardalos, P., Thoai, N.: Introduction to global optimization. Springer (1995)
26. Hubert, M., Rousseeuw, P., Vanden Branden, K.: Robpca: a new approach to robust principal component analysis. Technometrics **47**(1), 64–79 (2005)
27. Janak, S.L., Floudas, C.A.: Robust optimization: Mixed-integer linear programs. In: C.A. Floudas, P.M. Pardalos (eds.) Encyclopedia of Optimization, pp. 3331–3343. Springer US (2009)
28. Karmarkar, N.: A new polynomial-time algorithm for linear programming. Combinatorica **4**(4), 373–395 (1984)
29. Karush, W.: Minima of functions of several variables with inequalities as side constraints. MSc Thesis, Department of Mathematics. University of Chicago (1939)
30. Kim, S.J., Boyd, S.: A minimax theorem with applications to machine learning, signal processing, and finance. SIAM Journal on Optimization **19**(3), 1344–1367 (2008)
31. Kim, S.J., Magnani, A., Boyd, S.: Robust fisher discriminant analysis. Advances in Neural Information Processing Systems **18**, 659 (2006)
32. Kuhn, H., Tucker, A.: Nonlinear programming. In: Proceedings of the second Berkeley symposium on mathematical statistics and probability, vol. 481, p. 490. California (1951)
33. Lauer, F., Bloch, G.: Incorporating prior knowledge in support vector machines for classification: A review. Neurocomputing **71**(7–9), 1578–1594 (2008)
34. Lilis, G., Angelosante, D., Giannakis, G.: Sound Field Reproduction using the Lasso. Audio, Speech, and Language Processing, IEEE Transactions on **18**(8), 1902–1912 (2010)
35. Mangasarian, O., Street, W., Wolberg, W.: Breast cancer diagnosis and prognosis via linear programming. Operations Research **43**(4), 570–577 (1995)
36. McCarthy, J.: LISP 1.5 programmer's manual. The MIT Press (1965)
37. McCarthy, J., Minsky, M., Rochester, N., Shannon, C.: A proposal for the Dartmouth summer research project on artificial intelligence. AI MAGAZINE **27**(4), 12 (2006)
38. Minoux, M.: Robust linear programming with right-hand-side uncertainty, duality and applications. In: C.A. Floudas, P.M. Pardalos (eds.) Encyclopedia of Optimization, pp. 3317–3327. Springer US (2009)
39. Moore, G.: Cramming more components onto integrated circuits. Electronics **38**(8), 114–117 (1965)
40. Nielsen, J.: Nielsens law of Internet bandwidth. Online at http://www.useit.com/alertbox/980405.html (1998)
41. Olafsson, S., Li, X., Wu, S.: Operations research and data mining. European Journal of Operational Research **187**(3), 1429–1448 (2008)
42. Platt, J., Cristianini, N., Shawe-Taylor, J.: Large margin DAGs for multiclass classification. Advances in neural information processing systems **12**(3), 547–553 (2000)
43. Psychol, J.: 1. pearson k: On lines and planes of closest fit to systems of points in space. Phil Mag **6**(2), 559–572 (1901)
44. Rajaraman, A., Ullman, J.: Mining of massive datasets. Cambridge Univ Pr (2011)
45. Rao, C.: The utilization of multiple measurements in problems of biological classification. Journal of the Royal Statistical Society. Series B (Methodological) **10**(2), 159–203 (1948)
46. Rockafellar, R., Uryasev, S.: Optimization of conditional value-at-risk. Journal of risk **2**, 21–42 (2000)
47. Rosenblatt, F.: The Perceptron, a Perceiving and Recognizing Automaton Project Para. Cornell Aeronautical Laboratory (1957)
48. Santosa, B., Trafalis, T.: Robust multiclass kernel-based classifiers. Computational Optimization and Applications **38**(2), 261–279 (2007)

49. Schölkopf, B., Simard, P., Smola, A., Vapnik, V.: Prior knowledge in support vector kernels. Advances in neural information processing systems pp. 640–646 (1998)
50. Schölkopf, B., Smola, A.: Learning with Kernels. The MIT Press, Cambridge, Massachusetts (2002)
51. Shawe-Taylor, J., Cristianini, N.: Kernel methods for pattern analysis. Cambridge Univ Pr (2004)
52. Sim, M.: Approximations to robust conic optimization problems. In: C.A. Floudas, P.M. Pardalos (eds.) Encyclopedia of Optimization, pp. 90–96. Springer US (2009)
53. Sion, M.: On general minimax theorems. Pacific Journal of Mathematics $8(1)$, 171–176 (1958)
54. Tibshirani, R.: Regression shrinkage and selection via the lasso. Journal of the Royal Statistical Society. Series B (Methodological) pp. 267–288 (1996)
55. Tikhonov, A., Arsenin, V., John, F.: Solutions of ill-posed problems. Winston Washington, DC: (1977)
56. Trafalis, T., Alwazzi, S.: Robust optimization in support vector machine training with bounded errors. In: Neural Networks, 2003. Proceedings of the International Joint Conference on, vol. 3, pp. 2039–2042
57. Trafalis, T., Alwazzi, S.: Robust optimization in support vector machine training with bounded errors. In: Proceedings of the International Joint Conference On Neural Networks, Portland, Oregon, pp. 2039–2042. IEEE Press (2003)
58. Trafalis, T.B., Gilbert, R.C.: Robust classification and regression using support vector machines. European Journal of Operational Research $173(3)$, 893–909 (2006)
59. Tuy, H.: Robust global optimization. In: C.A. Floudas, P.M. Pardalos (eds.) Encyclopedia of Optimization, pp. 3314–3317. Springer US (2009)
60. Vapnik, V.: The Nature of Statistical Learning Theory. Springer-Verlag, New York (1995)
61. Walter, C.: Kryder's law. Scientific American $293(2)$, 32 (2005)
62. Xu, H., Caramanis, C., Mannor, S.: Robustness and Regularization of Support Vector Machines. Journal of Machine Learning Research 10, 1485–1510 (2009)
63. Xu, H., Caramanis, C., Mannor, S.: Robust regression and lasso. Information Theory, IEEE Transactions on $56(7)$, 3561–3574 (2010)